Digital Space in the Wake of Shrinking Traditional Media

Alfred Okoth Akwala
Technical University of Kenya, Kenya

Denish Ouko Otieno
Kenya Institute of Mass Communication, Kenya

Amos Moraro Marube
Technical University of Kenya, Kenya

IGI Global
Scientific Publishing
Publishing Tomorrow's Research Today

Vice President of Editorial Melissa Wagner
Director of Acquisitions Mikaela Felty
Director of Book Development Jocelynn Hessler
Production Manager Mike Brehm
Cover Design Jose Rosado

Published in the United States of America by
IGI Global Scientific Publishing
701 East Chocolate Avenue
Hershey, PA, 17033, USA
Tel: 717-533-8845 | Fax: 717-533-7115
Website: https://www.igi-global.com E-mail: cust@igi-global.com

Library of Congress Cataloging-in-Publication Data

Names: Akwala, Alfred O., 1976- editor. | Otieno, Denish Ouko, 1985- editor. | Marube, Amos Moraro, 1980- editor.
Title: Digital space in the wake of shrinking traditional media / edited by Alfred Akwala, Denish Otieno, Amos Marube.
Description: Hershey : Information Science Reference, 2026. | Includes bibliographical references and index. | Summary: "This book provides a comprehensive analysis of the changes occurring in the media and communication landscape due to the rise of digital media. The book examines the impact of digital media on traditional media sources such as newspapers, radio, and television, and explore the opportunities and challenges arising from this shift"-- Provided by publisher.
Identifiers: LCCN 2023038678 (print) | LCCN 2023038679 (ebook) | ISBN 9781668493564 (hardcover) | ISBN 9781668493601 (paperback) | ISBN 9781668493571 (ebook)
Subjects: LCSH: Journalism--History--21st century. | Journalism--Technological innovations. | Online journalism. | Digital media.
Classification: LCC PN4815.2 .D54 2023 (print) | LCC PN4815.2 (ebook) | DDC 070.9/051--dc23/eng/20231005
LC record available at https://lccn.loc.gov/2023038678
LC ebook record available at https://lccn.loc.gov/2023038679

British Cataloguing in Publication Data
A Cataloguing in Publication record for this book is available from the British Library.

Table of Contents

Detailed Table of Contents

The impact of digital media on public opinion and political debate is examined in this chapter. The significance of digital media in political communication and its effect on public opinion and political discourse are introduced at the outset of the chapter. It analyses the pros and cons of using social media for political campaigns. Further probes deeper into the proliferation of false information and propaganda in digital media, its effect on public opinion and political discourse, and the measures taken to counter it. It discusses the difficulties and ethical concerns of using digital media to encourage civic participation. This chapter provides a concise overview of the discussion and draws attention to the implications for future study and application.

This case study examines how political communication in India has evolved with the rise of digital technology, analyzing its impact on political campaigns, voter engagement, and public discourse. The study delves into the significance of social media platforms, digital messaging apps, and online news outlets in shaping the political landscape in the world's largest democracy.In recent years, India has witnessed a significant surge in internet and smartphone usage, leading to a transformation in the way political parties and leaders interact with their constituents. This study investigates the various tools and strategies employed by political actors to connect with voters and disseminate their messages in a digital-first environment.

Social TV is the use of social media and digital technologies to enhance the TV viewing experience. However, there is a lack of research on social TV in Nigeria. This chapter aims to fill this gap by providing a critical analysis of social TV and audience interaction in Nigeria with some preliminary findings. A modified UTAUT2 was used as the theoretical framework, integrating technology awareness and religiosity as additional predicting and moderating variables respectively. A survey among Nigerian university students reveals that performance expectancy, social influence, hedonic motivation, price value, habit and technology awareness significantly predict use intentions, with extrinsic religiosity strengthening the relationship between behavioural intention and actual usage. The chapter also explores the potential and implications of social TV for content production, civic engagement, and cultural diversity in Nigeria. The chapter concludes by summarising the findings and contributions, and highlighting the limitations and directions for future research.

The study was driven by the need to investigate the use of social media and its influence on how college students engage with each other; and their writing activities and projects. Writing is a key component of communication competence whether in interpersonal, workplace or audience engagement in media productions. Communication engagements are shaped by the medium and this is evident in a society centered around digital mediums that are continually changing and advancing in their features. The exponential growth of social media and the high rate of mobile phone penetration has led to platforms such as Facebook, Twitter, TikTok and WhatsApp being a key part in youths' daily communication; yet these social media platforms require users to express themselves using limited words complemented by memes and acronyms. This paper adopted a qualitative. Findings indicate that social media shapes the way students write as illustrated by the inconsistencies in their formal writing which is influenced by the informal writing, slang, emojis and acronyms used on social media.

The news media play an important role in informing the public on scientific research findings in health matters. In Kenya there has been unprecedented rise in the use of Social Network Sites more so among the youth. According to a report by SIMElab Africa, 2019, X is specifically used more by those aged 26 to 35 years in addition to being used by journalists and media organizations to disseminate information. Condit (2004) posits that there has been a mismatch between the expectations of the scientists and those of the reporter on disseminating scientific research findings which sometimes twists the messaging. The overall objective of the study was to establish framing of health research findings on X as a news media platform with focus given to The Daily Nation's Health and Science pull out '@DNHealthyNation'. The study found out that there is a mismatch between health reports on social network sites and original scientific health disseminations due to factors such as competition for readers' attention by journalists which leads to sensationalized report interpretations

Fake news is a problematic concept with multiple meanings for diverse persons. Journalists, politicians, academics, and even laypeople use it on a daily basis. However, more often than not they refer to different things. Therefore, in order to elaborate a more robust analysis of the term, it is important to clarify who utilises it and for what. Hence, the aim of this chapter is threefold: firstly, considering its content, this study proposes an updated typology of fake news that integrates humorous messages, disinformation, and unethical journalism practices. Secondly, it establishes a distinction between the message as such and its weaponization by populist political leaders. Thirdly, it reflects on the reasons why the audience constantly look for - and share - this kind of stories. In doing so, it draws on a comprehensive literature review.

With the development of Information Communication Technologies(ICTs), family engagement and to a larger extent social engagement has evolved from traditional media to digital media. Many countries have shifted from using agricultural extension officers to using of digital technology in accessing agriculture related information to enhance overall economic growth. Farmers in Kenya continue to face a multitude of problems to maximize crop and animal productivity. It is a fact that Kenya possesses valuable agricultural knowledge and expertise in addition to fertile soils. However, a wide information gap exists between research knowledge and the farmers. Farmers need timely expert advice to make them more productive and competitive.

The present chapter aimed to investigate the impact of information sources on anxiety related to potential discoveries by the James Webb Space Telescope (JWST). A sample of 1009 participants was surveyed, and descriptive and inferential statistical analysis was performed. Results indicated that 24.6% of the participants reported anxiety concerning JWST's discoveries, while 41.3% believed in the existence of extraterrestrial intelligence. The chapter revealed a significant correlation between information sources and anxiety towards JWST's discoveries (R = 0.268, R2 = 0.72, Adj R2 = 0.059, P < 0.05). It was discovered that library and online newspaper sources were the major contributors to such anxiety. The findings highlight the significance of information sources in shaping individuals' reactions to potential extraterrestrial discoveries. Therefore, the chapter recommends libraries and online newspapers to educate the public about JWST's findings, alleviating anxiety and promoting accurate information.

Foreword

As a tradigital and mentor to the three editors of this book, I am very pleased and honoured to write the foreword to this book. Our university's (Technical University of Kenya) mission is to contribute towards the advancement of society through research and innovation, and I am happy to hold in my hand a physical manifestation of this idea.

First of all, I would like to congratulate the Publishers and the Editors, Alfred Akwala (Technical University of Kenya, Kenya), Denish Otieno (Kenya Institute of Mass Communication, Kenya), and Amos Marube (Technical University of Kenya, Kenya) on their initiative in bringing together a wealth of information and expertise - between the covers of this book on the disruptions that the ubiquity of digital media has occasioned.

Previously, I mentioned that I am tradigital. This means I am still nostalgic about the past while faced with a new world order. Consequently, I am always seeking new information on how to deal with the current disrupted digital space. Thus, this book is instructive and will—I am sure—be of great interest to media students, practitioners, and policy interpreters who are keen on understanding media practice in the post-digital era.

Specifically, the book editors present complex and exciting papers that aid the reader in comprehending critical and theoretical perspectives on the evolution of media and communication. Secondly, in the post-digital era, the editors argue, the traditional media landscape is completely disrupted. Further, they factually claim that the post-digital presents challenges such as fake news, misinformation, disinformation, data privacy concerns and the political economy of the media industry.

This book will be a very welcome and invaluable resource and one that I can heartily recommend to all others who, in any way, are involved in the exciting and challenging world of the digital space against a backdrop of a shrinking legacy media. It will be instructive on critical enquiry, theoretical perspectives and philosophical positions.

Peter M. Matu

Faculty of Social Sciences and Technology, Technical University of Kenya, Kenya

Nairobi, August 2025

Preface

The digital era has dramatically redefined the media and communication landscape. Where once newspapers, radio, and television held uncontested sway over public discourse, we now find ourselves immersed in a sprawling, fast-evolving digital space. This reference volume, *Digital Space in the Wake of Shrinking Traditional Media*, is both a reflection on and a response to that transformation.

The impetus for this book lies in a series of fundamental shifts: the rapid decline of traditional media outlets, the rise of decentralized and user-driven platforms, and the profound implications—both promising and problematic—that these changes bring. As editors, we set out to explore and document this evolving environment not just as observers, but as participants in a larger conversation about the future of media, society, and democratic engagement.

Digital media has undeniably democratized information access, lowering barriers to entry and enabling voices from across the globe to be heard. It has spurred innovation, empowered creators, and fostered new communities of expression. Yet it has also introduced new threats: the spread of misinformation, deepening polarization, surveillance capitalism, and a troubling consolidation of power among tech giants.

This book aims to present a rigorous, balanced examination of these developments. We have drawn on diverse perspectives to offer a comprehensive overview of how digital media is reshaping traditional forms of communication, the economics behind them, and their role in public life. Our contributors interrogate pressing issues such as the transformation of journalism, the ethics of data use, and the cultural shifts driven by digital interaction.

Specifically, this volume is structured around four major themes:

1. The Economics of Digital Media – Exploring how digital platforms have disrupted legacy media business models and given rise to new ones, such as influencer economies, podcasting, and algorithm-driven content distribution.
2. Journalism and News in Transition – Investigating how the digital turn is affecting the quality, reliability, and practice of journalism, and what this means for informed citizenship in an age of information overload.
3. Social and Cultural Dimensions – Examining how digital media alters human interaction, identity, privacy norms, and creative expression, including the rise of online activism and participatory culture.
4. Ethics and Responsibility – Addressing the ethical challenges presented by digital technologies, from data security to algorithmic bias, while envisioning a more equitable and sustainable digital ecosystem.

Our intended audience is equally broad and diverse. Media professionals, journalists, and content creators will find this work valuable in understanding the shifting terrain of their industry. Academics and students will encounter insights relevant to ongoing debates in media studies, communication, and digital sociology. Entrepreneurs, marketers, and digital strategists will gain a clearer view of the forces shaping contemporary media markets.

More broadly, we hope this book serves all those seeking to engage critically and ethically with digital media—whether as users, creators, or regulators. As our information ecosystems continue to evolve, the need for thoughtful, well-informed engagement has never been more urgent.

We invite readers to use this volume not only as a source of knowledge, but as a springboard for further dialogue, reflection, and action in shaping a digital future that is just, inclusive, and sustainable.

ORGANIZATION OF THE BOOK

This edited volume presents a diverse and timely exploration of how digital media is reshaping public discourse, political engagement, education, journalism, and everyday life across different societies. The following chapters examine these themes through interdisciplinary lenses and case studies from various regions, particularly within the Global South. What emerges is a layered, critical conversation about digital transformation—its promises, its perils, and its profound implications for our collective future.

Chapter 1: The Role of Digital Media in Shaping Public Opinion and Political Discourse

Opening the volume is an analysis of the transformative role digital media plays in shaping public opinion and political communication. This chapter critically examines both the empowering and disruptive potentials of social media platforms in political contexts. It interrogates the use of digital tools in campaigns, the rise of misinformation and propaganda, and the ethical complexities of civic engagement in online spaces. The chapter also outlines existing countermeasures to digital manipulation and identifies urgent areas for future research and policy development.

Chapter 2: Changing Dynamics of Political Communication in Digital Era: A Case Study on India

Focusing on India, the world's largest democracy, this case study investigates the digital transformation of political communication. It traces how internet access and mobile technology have revolutionized the strategies political actors use to connect with constituents. The chapter details the use of social media, messaging apps, and online news platforms in electioneering, offering a comprehensive view of how the digital era has redefined political engagement and public discourse in a highly diverse and populous nation.

Chapter 3: Social TV and Nigerian Youth Culture Online: Literature, Religiosity and Preliminary Findings

This chapter explores the emerging phenomenon of social TV in Nigeria and its resonance with youth culture. Using a modified UTAUT2 framework, it analyzes how religiosity and technology awareness shape audience engagement. Preliminary findings from university students highlight key predictors of social TV adoption, while the chapter also examines broader implications for cultural production, civic dialogue, and audience behavior in a digitally mediated entertainment landscape.

Chapter 4: Social Media Usage and Writing Skills Competence Among College Students

Here, the relationship between students' social media habits and their writing competence is explored. Through qualitative inquiry, the chapter reveals how informal online communication—characterized by slang, emojis, and acronyms—affects students' formal writing tasks. It calls attention to the implications of digital expression for educational settings and raises important questions about how digital media shapes linguistic development and communication norms in academic contexts.

Chapter 5: Social Network Sites Audience Engagement on Health Reports: An Evaluation of Daily Nation's X Platform, @DNHealthyNation

This chapter turns to health communication in Kenya, evaluating how scientific research findings are framed and disseminated on social media, with a focus on the Daily Nation's @DNHealthyNation. It highlights a troubling gap between original research and its presentation to the public, driven in part by journalists' competition for audience attention. The analysis underscores the importance of responsible media framing, especially in public health, and urges improved collaboration between scientists and media professionals.

Chapter 6: Fake News for Whom? Discussing Types, Political Appropriation, and Consumption of False Contents

In this nuanced exploration of "fake news," the chapter critiques the term's conceptual vagueness and political weaponization. It introduces a typology of misinformation—from satirical content to deliberate disinformation—and examines how populist leaders exploit such content for political gain. Additionally, it investigates why audiences consume and circulate fake news, drawing on a comprehensive literature review to unpack the psychological and sociopolitical dynamics at play.

Chapter 7: Adoption of Information Communication Technologies in Enhancing Agricultural Productivity in Rural Kenya

Moving into the sphere of development, this chapter explores how ICTs are bridging the information gap in rural Kenyan agriculture. It documents the transition from traditional extension services to digital platforms that provide timely, expert advice to farmers. The chapter underscores the transformative potential of digital tools in enhancing agricultural productivity and rural livelihoods, while also pointing to the infrastructural and educational challenges that must be addressed.

Chapter 8: James Webb Space Telescope and Discovery Anxiety: Role of the Sources of Information

Concluding the volume is an investigation into how media sources influence public anxiety surrounding scientific discovery—in this case, the James Webb Space Telescope. The chapter draws on survey data to show a measurable link between information sources (like libraries and online news) and anxiety levels related to extraterrestrial findings. It calls for improved public communication strategies to balance excitement with scientific literacy and reduce unnecessary fear or misinformation.

Together, these chapters offer a comprehensive, critical lens through which to understand the intersection of digital media, society, and communication across global contexts. The volume underscores the urgency of interdisciplinary dialogue and responsible engagement as we continue to navigate the digital age.

CONCLUSION

As we draw this volume to a close, it is clear that digital media is not simply a tool or platform—it is a transformative force reshaping the foundations of communication, politics, education, culture, and public life. Across the chapters presented in this reference book, we have brought together a range of perspectives that interrogate how individuals, institutions, and societies are adapting to this ongoing digital shift, and what is at stake in that adaptation.

From the use of social media in political discourse and influencer culture, to its influence on journalistic integrity, academic writing, agricultural productivity, and public anxiety about science, the breadth of inquiry captured here reveals the layered complexity of our digital realities. The cases and studies presented—many rooted in the Global South—highlight not only the global reach of digital transformation, but also its context-specific challenges and opportunities. These insights challenge the often-Western-centric lens through which digital media is analyzed, offering a richer, more inclusive understanding of how digital media operates in diverse cultural and socio-political environments.

This volume does not attempt to provide definitive answers. Instead, it invites ongoing reflection and critical inquiry. We hope that scholars, media practitioners, students, policy makers, and engaged citizens will find in these pages a foundation for deeper thinking, ethical consideration, and informed action.

Digital media will continue to evolve, often faster than our ability to fully comprehend its implications. Yet what remains essential is a commitment to media literacy, transparency, inclusivity, and accountability in both the production and consumption of digital content. As editors, we are hopeful that this book contributes meaningfully to that commitment, fostering a more equitable, participatory, and critically engaged digital society.

Chapter 1
The Role of Digital Media in Shaping Public Opinion and Political Discourse

Surjit Singha
https://orcid.org/0000-0002-5730-8677
Kristu Jayanti University, India

Ranjit Singha
https://orcid.org/0000-0002-3541-8752
Christ University, India

R. Sreethi Rebeka
https://orcid.org/0000-0003-3659-8027
Christ University, India

ABSTRACT

The impact of digital media on public opinion and political debate is examined in this chapter. The significance of digital media in political communication and its effect on public opinion and political discourse are introduced at the outset of the chapter. It analyses the pros and cons of using social media for political campaigns. Further probes deeper into

DOI: 10.4018/978-1-6684-9356-4.ch001

the proliferation of false information and propaganda in digital media, its effect on public opinion and political discourse, and the measures taken to counter it. It discusses the difficulties and ethical concerns of using digital media to encourage civic participation. This chapter provides a concise overview of the discussion and draws attention to the implications for future study and application.

INTRODUCTION

Digital media has revolutionised political communication, influencing public opinion and political discourse in previously inconceivable ways. Notably, social media has become an indispensable instrument in political campaigns. The proliferation of disinformation and false news has highlighted the need for the ethical and responsible use of digital media in political communication. Moreover, digital media can potentially promote civic engagement and participation in political processes. This chapter focuses on the influence of social media, the proliferation of disinformation and fake news, and the potential of digital media to foster civic involvement. It investigates digital media's role in shaping public opinion and political debate (López-López et al., 2023; Oparaugo, 2021).

Digital media have emerged as a significant force in modern society's public opinion and political discourse. It has revolutionised information dissemination, consumption, and sharing, giving birth to a culture where individuals can actively engage with political content and convey their opinions. Digital media's significance rests in its capacity to amplify voices, facilitate the dissemination of information, and provide forums for public discussion and debate (Chadwick, 2013).

Digital media platforms, such as social networking sites, blogs, and online news publications, provide a democratised space for individuals to voice their opinions, share information, and participate in political discussions. These platforms have a broad reach and allow individuals from disparate backgrounds to contribute to the public discourse regardless of geographical or social barriers. This extensive participation can challenge traditional power structures, make it possible for marginalised voices to be heard, and cultivate a more inclusive political climate (Bernacki et al., 2021).

Digital media play a crucial role in setting agendas and shaping public opinion. The pace and accessibility of online news and social media have altered how individuals ingest and interpret information. Through digital channels, news travels rapidly, and individuals can access various perspectives and alternative sources beyond traditional media outlets. This abundance of information sources enables individuals to craft their narratives and contest dominant discourses. Furthermore, digital media promotes political mobilisation and activism. Social media platforms have played a crucial role in organising grassroots movements, protests, and advocacy campaigns, empowering individuals to mobilise around political causes and effect change. The ubiquitous nature of online content can generate substantial public interest and pressure, influencing political agendas and decisions (Kumar, 2023). Kwon et al. (2016) conducted a study investigating the role of social media rumours as improvised and expressive public opinion during uncertain sociopolitical situations. Using a theoretical framework from early rumour studies, the researchers analysed Twitter messages circulated during the 2013 Korean saber-rattling. Similar to institutionalised opinion polling, non-rumour narratives focused predominantly on policy-level responses to the threat situation, as disclosed by the study.

In contrast, rumours reflected the hegemonic tensions between anti-leftist political sentiments and counteractive accounts rather than a concern for official responses. These rumours frequently incorporated humour, speculation, or desires as a coping mechanism. The study highlighted the significance of scrutinising online rumours to understand how collective memories interact with present uncertainties, thereby shaping public opinions and sentiments (Kwon et al., 2016). By thoroughly analysing the content and semantic networks within Twitter messages, the study cast light on the dynamic nature of public opinion and the underlying social and political tensions that influence public discourse during uncertain times. Digital rumors spread quickly and widely due to the interconnected nature of social media platforms. This rapid spread can cause significant societal impacts in a short period (Chau et al., 2022 ; Du Chau et al., 2022; Hindawy, 2021). The spread of rumors is influenced by user behaviors, including the tendency to share content that aligns with their beliefs and the role of echo chambers in reinforcing these beliefs(Du Chau et al., 2022; Ray & George, 2019)

BACKGROUND

The relationship between digital media and politics dates back to the advent of the internet in the late 20th century. The development of the World Wide Web and the proliferation of personal computers enabled individuals to access and share information globally. Initially, digital media primarily supplemented traditional media outlets, serving as an alternative source of information. However, with the emergence of blogs in the late 1990s, a significant transition in the digital media landscape occurred, reshaping its influence on politics. Blogs allowed individuals to express their opinions, comment on political events, and engage in discussions, effectively challenging the mainstream media's traditional gatekeeping role. This shift fostered a greater diversity of voices and perspectives in political discourse (De Zúñiga & Chen, 2019; Oser & Boulianne, 2020).

The mid-2000s witnessed a further transformation with the advent of social media platforms such as Facebook and Twitter, which amplified digital media's impact on politics. These platforms facilitated rapid information dissemination, real-time interactions, and the formation of politically focused online communities. Political actors—including candidates, parties, and interest groups—began leveraging social media as a strategic tool for communication, mobilization, and persuasion. Digital media has since revolutionized political campaigns, enabling candidates to engage directly with voters. Online news outlets and citizen journalists have also challenged the dominance of traditional media organizations, empowering individuals to shape narratives and challenge established political authorities (Kaur & Kaur, 2013). Digital media has been shown to increase political participation. For example, the use of social media for political purposes positively influences political engagement, as seen in the 2017 Legislative Assembly Elections in Punjab, India (Kaur & Verma, 2018). Similarly, digital media use is associated with broader political participation in the United States (Copeland & Bimber, 2014). Social media platforms like

Facebook and Twitter have empowered youth to actively shape political preferences and opinions, particularly in Malaysia (Sapar & Hussin, 2024).

While digital media offers significant opportunities, its pervasive use in politics also presents challenges. Misinformation, filter bubbles, and the erosion of trust in traditional institutions have become pressing concerns. This chapter examines the role and influence of digital media on public opinion and political discourse by analyzing how digital platforms shape the public agenda, impact political narratives, and influence political participation. Furthermore, it explores the implications of digital media on democratic processes, particularly in fostering civic engagement, amplifying marginalized voices, and enhancing government transparency. Ethical considerations and responsibilities related to digital media's role in politics will also be addressed. The spread of misinformation and fake news is a significant ethical concern in digital media. Ensuring the accuracy and truthfulness of information is crucial to maintaining public trust and the integrity of democratic processes (Banciu & Cirnu, 2022; Ribeiro, 2024). Ethical digital practices involve promoting transparency, honesty, and accountability in communication to prevent the manipulation of public opinion(Banciu & Cirnu, 2022). Privacy breaches and the misuse of personal data are critical issues. Digital media professionals must prioritize the protection of users' privacy and ensure data security to prevent unauthorized access and misuse(Kgatla, 2024; Patalauskaitė, 2024).

Ultimately, this chapter aims to provide a comprehensive understanding of digital media's role in shaping public opinion and political discourse, critically assessing its impact on democracy and political processes.

RESEARCH PROBLEM

The increasing reliance on digital media as a primary source of political information has significantly altered how public opinion is formed and how political discourse unfolds. While digital media platforms enable rapid information sharing and political engagement, they also contribute to the spread of misinformation, the reinforcement of ideological echo chambers, and the manipulation of public perception. Additionally, concerns regarding algorithmic biases, data privacy, and the ethical responsibilities of digital media platforms in political communication remain unresolved. This research seeks to explore the dual role of digital media as both an enabler of democratic participation and a potential threat to political integrity, assessing its broader implications on governance, public trust, and civic engagement.

RESEARCH OBJECTIVES

1. To examine how digital media influences public opinion formation and political discourse.
2. To analyze the role of social media platforms in shaping political narratives and influencing voter behavior.
3. To investigate the impact of digital media on political participation, including mobilization and civic engagement.
4. To explore the challenges posed by misinformation, filter bubbles, and declining trust in traditional institutions due to digital media.
5. To assess the ethical responsibilities and regulatory considerations related to the use of digital media in politics.

RESEARCH QUESTIONS

1. How does digital media influence public opinion and political discourse?
2. In what ways do social media platforms shape political narratives and voter behavior?
3. What role does digital media play in increasing or hindering political participation and civic engagement?
4. How do misinformation and algorithm-driven content curation impact public trust in political institutions?
5. What ethical and regulatory considerations should be implemented to ensure responsible use of digital media in political communication?

THEORETICAL FOUNDATIONS

Selective Exposure

People seek information consistent with their beliefs and values, averting contradictory perspectives. This particular exposure can contribute to the echo chamber effect, in which individuals are surrounded by individuals with similar viewpoints, thereby reinforcing their own biases (Theocharis et al., 2016).

Social Influence

According to the theory of social influence, individuals are influenced by their social networks and the views of others. When misinformation is disseminated by credible sources or within social networks, individuals may be more likely to accept and spread it without evaluating its veracity critically (Davis et al., 2017).

Cognitive Biases

Misinformation is partially disseminated and accepted due to cognitive fallacies such as confirmation bias and availability heuristics. Confirmation bias causes individuals to interpret information in a manner that confirms their preexisting beliefs, whereas the availability heuristic influences judgement based on the readily available information (Nickerson, 1998).

Information Overload

The digital age's abundance of information can overwhelm individuals, making distinguishing between accurate and false information difficult. Information excess may result in rash decisions and a greater propensity to accept and spread incorrect information.

Networked Public Sphere

According to this theory, digital media have transformed the public sphere by facilitating decentralised, networked conversations. It highlights the role of online platforms in facilitating deliberation, information exchange, and collective action, thereby broadening the scope and inclusiveness of political discourse.

Agenda-Setting Theory

Digital media is vital in moulding public perception and priorities by emphasising specific issues and influencing which topics receive attention. Digital media platforms' algorithms and curation mechanisms can shape the agenda and political discourse.

Technological Determinism

Technological determinism proposes that digital media technologies possess inherent qualities and affordances that influence political discourse. The characteristics and layout of digital media platforms impact how information is presented, ingested, and shared, thereby affecting the nature and dynamics of political discussions.

Spiral of Silence Theory

This theory proposes that digital media can influence public opinion by fostering an environment where individuals feel more secure expressing their thoughts when they believe they are in the majority. Through the visibility and amplification of particular viewpoints, digital media platforms can affect the perception of public opinion, potentially suppressing dissenting opinions.

Selective Exposure Theory

The argument suggests that individuals seek out and consume information that aligns with their beliefs and opinions. With the advent of digital media and its personalised content algorithms, there is a concern that this phenomenon can contribute to the polarisation of public opinion. By providing users with content that confirms their preconceived notions, digital media has the potential to reinforce existing biases and further divide society.

Agenda-Setting Theory

Agenda-setting theory suggests that digital media contributes to the formation of public opinion by influencing the perceived salience and significance of various issues. The topics and narratives highlighted in digital media can influence public discourse, influencing the perceptions and priorities of the public.

CONCEPTUAL FRAMEWORK

The conceptual framework for understanding the role of digital media in shaping political discourse integrates multiple theoretical foundations that explain how information is disseminated, consumed, and influences public opinion. The framework is structured around key interrelated components: Selective Exposure, Social Influence, Cognitive Biases, Information Overload, Networked Public Sphere, Agenda-Setting, Technological Determinism, and the Spiral of Silence. These elements interact within digital media environments, shaping the nature and inclusiveness of political discussions.

Information Consumption and Selective Exposure

According to Selective Exposure Theory, individuals actively seek and consume information that aligns with their preexisting beliefs while avoiding contradictory perspectives. This phenomenon is exacerbated by algorithm-driven content curation on digital media platforms, leading to the formation of echo chambers that reinforce ideological biases and contribute to political polarization. Cognitive biases, such as confirmation bias and availability heuristics, further shape the way individuals interpret and disseminate political information. These biases influence users to favor information that aligns with their worldview while dismissing counterarguments, affecting public discourse's diversity and objectivity.

Social Influence and the Spread of Information

Digital media fosters an interconnected environment where individuals are significantly influenced by their social networks. Social Influence Theory posits that people are more likely to accept and propagate information when it comes from credible sources within their social circles. In an era of rapid information exchange, misinformation spreads quickly when accepted within echo chambers, reducing the likelihood of critical evaluation. The Spiral of Silence Theory adds another layer to this dynamic, suggesting that individuals may refrain from expressing dissenting opinions if they perceive their views as being in the minority, further narrowing the diversity of perspectives in online political discourse (Kelman, 1958).

Traditional models often use a single influence mechanism. However, newer models recognize multiple influence mechanisms and types of relationships among individuals, enhancing the analysis of influence effects in various social context (Wicker & Doyle, 2012).

Digital Media's Role in Agenda-Setting and Public Discourse

The Agenda-Setting Theory underscores the significant role digital media plays in shaping public perception and priorities. Through algorithm-driven content prioritization and media coverage, specific issues gain prominence in political discussions, influencing public opinion and policy discourse. This agenda-setting function is intertwined with the concept of the Networked Public Sphere, which highlights how digital media decentralizes discussions, enabling broader participation and grassroots activism. However, while digital platforms provide a space for deliberation, they also reinforce dominant narratives, limiting exposure to alternative viewpoints.

Technological Determinism and the Structuring of Political Discourse

The Technological Determinism perspective asserts that the inherent design and affordances of digital media platforms shape the way information is presented, consumed, and shared. Features such as instantaneous communication, virality, and interactive engagement influence the dynamics of political discourse, often amplifying sensationalist content while overshadowing nuanced discussions. The design of digital platforms, through recommendation algorithms and engagement-driven metrics, fosters an environment where certain narratives gain traction while others are suppressed, reinforcing selective exposure and polarization.

Implications for Political Engagement and Collective Action

The convergence of these theories within digital media environments has profound implications for political engagement and collective action. The increasing prevalence of information overload makes it difficult for individuals to critically assess the credibility of sources, leading to reliance on heuristics and social validation when processing information. In turn, this affects the quality of civic engagement and the effectiveness of digital activism. While digital media has expanded access to political discourse, it has also introduced new challenges in ensuring informed, balanced, and inclusive participation.

This conceptual framework highlights the interplay between individual cognitive processes, social influences, technological affordances, and media agenda-setting in shaping political discourse. While digital media democratizes access to political discussions, it also fosters polarization, misinformation, and selective exposure. Addressing these challenges requires promoting media literacy, enhancing algorithmic transparency, and fostering diverse and respectful discussions in digital spaces. By understanding the theoretical foundations that underpin digital political engagement, policymakers, media professionals, and citizens can work toward creating a more informed and inclusive digital public sphere.

SOCIAL MEDIA IN POLITICAL CAMPAIGNS

"Social media" refers to a broad category of Internet-based platforms and apps facilitating user communication and collaboration. Typically, these platforms enable users to establish profiles, interact with others, and engage in various forms of communication, including sharing posts, remarks, images, and videos.

Social media is crucial in reaching and engaging electors during political campaigns. It provides a channel for direct, interactive communication between political candidates, parties, and their supporters. Candidates can utilise social media platforms to communicate their policy stances, campaign updates and engage in dialogue with voters. In addition, campaigns can target specific demographics and tailor their messages using social media. According to Dimitrova and Matthes (2018), the future of political campaigns is intricately intertwined with social media's content, applications, and effects. They emphasise the significance of future researchers actively addressing these challenges. By confronting the complexities of social media's role in political campaigns, researchers can contribute to a more comprehensive understanding of this dynamic landscape and its impact on the political process. To effectively investigate and navigate this new digital era, it is essential to recognise the changing character of political communication and adapt theoretical and methodological approaches accordingly.

Several political campaigns have utilised social media to their advantage. The 2008 presidential campaign of Barack Obama is a prominent illustration. Obama's campaign used social media platforms, namely Facebook and Twitter, to effectively mobilise supporters, raise funds, and disseminate campaign messages. At the time, the campaign's innovative use of social media enabled unprecedented levels of grassroots engagement and online activism.

Donald Trump's 2016 presidential campaign is another notable example. Trump's campaign utilised Twitter extensively, allowing him to bypass traditional media channels and communicate directly with his supporters. His provocative and attention-getting tweets garnered extensive media coverage and shaped the campaign's narrative. The social media campaign strategy mobilised a loyal constituency and contributed to Trump's electoral victory.

SOCIAL MEDIA IN POLITICS

Social media has transformed political campaigns by providing a wide reach, enabling candidates to engage with diverse audiences across geographical and demographic lines. With targeted messaging, campaigns can tailor their communication strategies based on users' interests and online behaviors, enhancing message effectiveness. The real-time communication capabilities of social media foster direct interaction between candidates and voters, promoting accessibility and responsiveness. Additionally, the amplification of messages through shares and viral content increases the visibility and impact of campaign messages, while grassroots engagement allows supporters to mobilize communities, organize events, and expand participation in political movements. However, despite these advantages, social media also presents significant challenges. The spread of misinformation can distort public opinion and undermine democratic discourse. Filter bubbles and polarization, driven by algorithmic content curation, reinforce ideological divisions and limit exposure to diverse perspectives. Concerns about privacy and security persist due to the vast amounts of personal data collected and the potential for platform manipulation by malicious actors. Furthermore, social media can lead to superficial engagement, where individuals participate in political discussions without meaningful offline action, often referred to as "slacktivism." Lastly, the unequal access and digital divide highlight disparities in internet connectivity and socioeconomic status, restricting opportunities for political participation among marginalized groups. While social media offers immense potential for political engagement, addressing its drawbacks is crucial for fostering a more informed and inclusive democratic process. Socioeconomic status significantly affects internet access, with lower-income and less-educated individuals having reduced access to high-speed internet and digital devices (Abubakari et al., 2020; Willis & Tranter, 2006). This disparity is evident in both developed and developing countries, affecting marginalized communities' ability to participate in digital political activities (Rodríguez-Martínez & Lozano, 2021; Tirado-Morueta et al., 2017).

Social media platforms like Twitter and Facebook have transformed political engagement by providing new avenues for citizens to participate in political discussions, share opinions, and mobilize support for causes or candidates (Getachew & Beshah, 2019; Komodromos, 2015).

ENSURING ETHICAL INTEGRITY IN SOCIAL MEDIA-DRIVEN

The rise of social media as a powerful tool in political campaigns has transformed the way candidates interact with voters, shape public discourse, and mobilize support. However, this digital shift brings significant ethical responsibilities that must be upheld to maintain fairness, transparency, and trust in the democratic process. Ethical concerns such as transparency, privacy, accountability, and respectful discourse are essential in ensuring that social media serves as a platform for constructive political engagement rather than misinformation, manipulation, and polarization (Kgatla, 2024).

One of the fundamental ethical principles in social media-driven political campaigns is transparency. Campaigns must be open about the sources of their messaging, particularly when using sponsored content, targeted advertisements, or algorithm-driven strategies to reach specific demographics. Political advertisements should be clearly labeled, and the sources of funding should be disclosed to ensure voters can make informed decisions. Misleading tactics, such as fabricated endorsements, deepfake videos, and manipulated statistics, must be strictly avoided, as they can distort public perception and undermine the integrity of electoral processes. Transparency in political advertising involves clear disclosure of sponsorship. This means that political ads should explicitly state who is funding them. This practice helps voters understand the source of the information and assess its credibility (Jost et al., 2023; Wood & Ravel, 2018).

Another critical concern is privacy and data protection. Political campaigns often collect vast amounts of personal data to refine their messaging and tailor content to specific voter groups (Sheth, 2019). While data-driven strategies can enhance voter outreach, they must be conducted in compliance with data protection regulations. Campaigns should obtain user consent before collecting or utilizing personal information, ensuring that sensitive data is handled responsibly and not exploited for manipulative purposes. The misuse of data, as seen in past political scandals, raises ethical questions about voter profiling, microtargeting, and the potential for digital surveillance.

Accountability is another key ethical responsibility in social media-driven campaigns. Political candidates and campaign teams must take responsibility for the accuracy of the content they share (Alalawi et al., 2023). The rapid spread of misinformation and fake news on digital platforms poses a significant threat to public trust in political institutions. Campaigns should implement fact-checking mechanisms, promptly correct any inaccuracies, and refrain from disseminating unverified information. The deliberate spread of false claims, conspiracy theories, or misleading narratives can distort electoral outcomes and fuel public cynicism toward political participation.

Additionally, fostering civility and respectful discourse in online political discussions is crucial. Social media has become a breeding ground for hate speech, harassment, and personal attacks, which can discourage meaningful political dialogue. Political campaigns should actively promote respectful engagement, discourage divisive rhetoric, and denounce harmful online behavior (Trindade, 2023). Moderation policies should be in place to prevent toxic interactions, ensuring that social media platforms remain spaces for constructive political conversations rather than hostile battlegrounds.

Furthermore, campaigns must avoid manipulative tactics and voter suppression. The ethical use of social media in politics requires a commitment to honesty and fairness. Tactics such as creating fake accounts, employing bots to amplify support, or spreading disinformation to confuse or discourage voters violate the principles of ethical campaigning. Social media strategies should empower voters rather than manipulate their perceptions or suppress their participation.

Upholding these ethical standards is imperative to safeguarding democracy in the digital age. As social media continues to play an influential role in shaping political discourse, political campaigns must prioritize ethical integrity, ensuring that digital platforms are used responsibly to foster informed, inclusive, and fair elections.

THE RISE OF MISINFORMATION AND FAKE NEWS

The prevalence of misinformation and false news has increased in the digital media landscape. The viral nature of online platforms and the simplicity of creating and distributing content have contributed to spreading false or misleading information. Misinformation refers to inaccurate or false information frequently applied unintentionally, whereas fake news refers to intentionally fabricated or deceptive news stories.

Misinformation and false news are widely disseminated through digital media platforms, especially social media. Within specific user networks, the algorithmic nature of social media feeds and the prevalence of filter bubbles can amplify the spread of false information. In addition, the absence of gatekeeping and fact-checking mechanisms on social media platforms permits the space of incorrect information unchecked.

Cognitive biases, such as confirmation bias and availability heuristics, increase individuals' propensity to believe and share information that accords with their preexisting beliefs or is readily available. The swift nature of online communication also contributes to the unchecked spread of information. In their Chicago, Philadelphia, and Miami study, Wagner and Boczkowski (2019) conducted in-depth interviews to investigate people's perceptions of news reporting. The findings revealed a prevailing negative view among participants regarding the quality of news reporting. Additionally, there was a significant distrust towards news disseminated through social media platforms. Participants also expressed concerns about the information consumption habits of others. To counteract these perceptions, participants employed various strategies. They relied on traditional fact-based media sources, recognising their credibility and reliability. Personal experience and knowledge were also valued as essential factors in evaluating the integrity of the news. Participants highlighted the significance of encountering information repetitively across different channels, as it reinforced their confidence in its accuracy. Participants actively sought out and consumed information from cross-ideological sources. This approach allowed them to understand other perspectives and avoid potential biases comprehensively. Fact-checking emerged as a crucial practice, with participants emphasising verifying information before accepting it as accurate. News consumption is enduring a ritualised transition, with social media serving as a portal to the news. Journalists may respond to audience demands by delivering more fact-based content to combat misinformation and false news concerns.

THE IMPACT OF MISINFORMATION AND FAKE NEWS

Misinformation and fake news have become pervasive challenges in the digital age, significantly influencing public opinion and shaping political discourse. The widespread dissemination of false information through social media platforms, online news sources, and even mainstream media has profound implications for trust, polarization, public perception, and democratic processes. The ease with which misinformation spreads—often faster than factual information—has made it increasingly difficult for individuals to distinguish between credible news and deceptive narratives, ultimately affecting their ability to make informed political decisions (Ramesh et al., 2024).

Misinformation erodes trust in media, government institutions, and democratic processes. Repeated exposure to false content fosters skepticism, diminishing confidence in traditional journalism and expert opinions. This distrust weakens public faith in elections, governance, and policymaking, creating an environment where misinformation thrives and destabilizes democracy (De Zúñiga & Chen, 2019; Ecker et al., 2024).

Misinformation also fuels political polarization by reinforcing ideological biases through algorithm-driven echo chambers. Social media platforms amplify content aligned with users' beliefs, limiting exposure to diverse perspectives and deepening societal divisions. As individuals become entrenched in their views, meaningful debate and consensus-building decline, weakening social cohesion.

Beyond polarization, misinformation is often weaponized by state-sponsored groups, political operatives, and special interest organizations to manipulate public opinion. False narratives discredit opponents, fabricate scandals, and incite fear, influencing voter behavior and altering election outcomes. Such tactics distort reality, impair informed decision-making, and undermine democratic integrity.

Misinformation can also disrupt democratic processes by spreading falsehoods about elections, leading to voter suppression and confusion. Conspiracy theories delegitimize results, eroding trust in political institutions. Addressing this issue requires media literacy, fact-checking, and regulatory measures. Collaborative efforts among governments, tech companies, journalists, and citizens are essential to safeguarding democratic values and ensuring political discourse remains informed and credible.

EFFORTS TO COMBAT MISINFORMATION AND FAKE NEWS

In response to the growing threat of misinformation and fake news, various efforts have been undertaken by fact-checking organizations, social media platforms, educators, news agencies, and individual users to mitigate its spread and minimize its impact on public opinion and democratic processes. Given the rapid dissemination of false information online, these collective actions are essential in fostering an informed society and ensuring the integrity of political discourse.

One of the most significant efforts in combating misinformation is the role of fact-checking organizations, which work diligently to authenticate and debunk false claims. These organizations, such as PolitiFact, Snopes, and FactCheck.org, assess the credibility of news articles, public statements, and viral social media posts. By conducting thorough research and using reliable sources, fact-checkers help counter misinformation by providing verified and evidence-based information to the public. Many news agencies have also established their own fact-checking departments to scrutinize political statements, debunk conspiracy theories, and provide accurate reporting(Ecker et al., 2024). This verification process is essential in preventing the widespread acceptance of misleading narratives.

Another key approach to tackling misinformation is the implementation of platform interventions by social media companies. In recent years, platforms such as Facebook, Twitter, and YouTube have introduced various measures to curb the spread of false information. These include prioritizing authoritative sources in search results, flagging or labeling disputed content, and reducing the visibility of misleading posts. For instance, during election periods and public health crises, platforms have directed users to official sources of information, such as government websites and reputable news agencies, to prevent misinformation from influencing public behavior. Additionally, many platforms have partnered with fact-checking organizations to provide real-time verification of viral claims and warn users about misleading content. However, while these measures have been helpful, they also raise concerns about censorship and the potential for bias in content moderation, necessitating ongoing evaluation and improvement.

Beyond platform interventions, media literacy and education play a crucial role in equipping individuals with the skills to critically assess the information they consume. Media literacy initiatives are designed to teach individuals how to evaluate sources, verify facts, and recognize biased or manipulated content. Schools, universities, and non-governmental organizations (NGOs) have developed educational programs that help students and citizens navigate the complexities of the digital media landscape. By fostering critical thinking and analytical skills, media literacy efforts empower individuals to question the credibility of news and avoid falling prey to misinformation. Governments and educational institutions worldwide have recognized the importance of integrating media literacy into curricula to promote responsible media consumption (De Zúñiga & Chen, 2019; Ecker et al., 2024).

In addition to educational efforts, collaboration between social media platforms and news organizations has been instrumental in ensuring the availability of accurate information. By working together, these entities can support high-quality journalism, amplify verified news, and counter the spread of misleading narratives. Some platforms provide financial support to credible news agencies, helping them sustain investigative journalism and fact-checking initiatives. Moreover, these partnerships encourage the dissemination of trustworthy news by leveraging digital tools to reach wider audiences.

Finally, user empowerment remains a vital aspect of combating misinformation. Encouraging individuals to take an active role in identifying and reporting false information can significantly reduce the impact of misinformation on social and political discourse. Users can contribute by fact-checking before sharing content, reporting false or misleading posts, and promoting credible sources (De Zúñiga & Chen, 2019; Ecker et al., 2024). Responsible sharing practices, combined with a culture of verification, can help create a more informed and resilient digital community.

Despite significant progress in addressing misinformation, it remains a persistent challenge that requires ongoing efforts from multiple stakeholders, including technology companies, governments, media organizations, and the general public. The fight against fake news is a complex and evolving battle, demanding a collaborative and multifaceted approach to safeguard democracy, protect public trust, and ensure the integrity of information in the digital age.

THE POTENTIAL FOR DIGITAL MEDIA

Digital media has emerged as a powerful tool for fostering civic engagement, providing individuals with accessible platforms to participate in political processes, share their opinions, and mobilize for collective action. By enabling real-time access to information, facilitating dialogue, and expanding opportunities for participation, digital media has transformed how people interact with governance, activism, and social change. As digital platforms continue to evolve, their potential to enhance civic engagement grows, ensuring broader and more inclusive participation in democratic processes.

One of the most significant ways digital media promotes civic engagement is through accessible information. Online platforms, including news websites, government portals, and social media, provide instantaneous access to vast amounts of data on governance, public issues, and policy-making. This accessibility empowers individuals to stay informed about current affairs, evaluate different perspectives, and engage in meaningful discussions. Digital media allows citizens to fact-check statements made by politicians, track government activities, and follow legislative developments, ensuring that they can make informed decisions about political matters. Moreover, digital tools such as interactive data visualizations, podcasts, and explainer videos enhance public understanding of complex topics, making civic knowledge more engaging and accessible.

Beyond providing information, digital media serves as a space for amplifying voices, allowing individuals to express their opinions and advocate for issues they care about. Social networking sites, blogs, and online forums provide platforms where individuals can participate in discussions, share their perspectives, and connect with others who have similar interests. This digital discourse nurtures a sense of community and collective action, helping marginalized voices gain visibility. Unlike traditional media, which is often controlled by a select few, digital platforms democratize information dissemination, enabling ordinary citizens to shape public narratives. Movements like #MeToo and #BlackLivesMatter exemplify how digital media can amplify social issues, sparking global conversations and influencing policy changes.

Another critical aspect of digital media's impact on civic engagement is its role in grassroots mobilization. Activists and organizations leverage digital platforms to organize protests, promote petitions, and rally support for social causes. Hashtag campaigns, viral videos, and crowdsourcing initiatives have been instrumental in spreading awareness and uniting people around common objectives. Social media facilitates rapid coordination of events, such as climate strikes and human rights demonstrations, allowing activists to bypass traditional organizational barriers. Platforms like Twitter and Facebook enable real-time updates, while messaging apps such as WhatsApp and Telegram help organizers communicate securely and efficiently. By lowering the cost and effort required for political engagement, digital media enables grassroots movements to gain momentum and effect change at local, national, and global levels.

Furthermore, digital media expands opportunities for political participation beyond traditional channels. Citizens can interact directly with politicians, political parties, and government officials through comments, direct messages, and live-streamed events. Online town halls, Q&A sessions, and public forums allow elected representatives to engage with constituents in real time, fostering transparency and accountability. Digital media also enhances electoral participation by providing information on voter registration, polling locations, and election procedures. Some governments have even explored online voting systems, making it easier for citizens to cast their ballots. Additionally, citizen journalism, where individuals document and share news from their communities, has grown in prominence, holding authorities accountable and bridging gaps in mainstream media coverage.

As digital media continues to shape civic engagement, it is essential to address challenges such as misinformation, digital divides, and online polarization. However, with responsible use and continued innovation, digital platforms can strengthen democracy by empowering individuals to participate actively in political and social processes. By making information more accessible, amplifying diverse voices, facilitating grassroots mobilization, and enhancing political participation, digital media plays a crucial role in fostering an engaged and informed citizenry in the digital age.

EXAMPLES OF SUCCESSFUL DIGITAL MEDIA

Numerous successful digital media campaigns have promoted civic engagement. Here are a few notable examples:

The Black Lives Matter Movement

Black Lives Matter utilised social media to bring attention to racial injustices and police brutality. Hashtags like #BlackLivesMatter and

#SayHerName attracted widespread attention, spurring online discussions and mobilising offline protests and advocacy efforts.

The ALS Ice Bucket Challenge

This viral campaign raised awareness and funds for research into amyotrophic lateral sclerosis (ALS). Participants would film themselves pouring a pail of frigid water over their heads and then dare others to do the same. The social media campaign went viral, attracting millions of participants and generating substantial donations.

The Arab Spring

Digital media platforms propelled the Arab Spring uprisings in several Middle Eastern nations. Social media played a crucial role in organising protests, disseminating information, and mobilising citizens, resulting in significant regional political shifts.

CHALLENGES TO PROMOTING CIVIC ENGAGEMENT

While digital media has become a transformative force in promoting civic engagement, it also faces several challenges that can hinder its effectiveness in fostering meaningful participation in political and social discourse. Despite its ability to connect people, provide information, and mobilize grassroots movements, digital media is not without its drawbacks. Issues such as the digital divide, filter bubbles, misinformation, and online harassment pose significant threats to the inclusive and democratic nature of digital engagement (De Zúñiga & Chen, 2019). Addressing these obstacles is essential to ensuring that digital media fulfills its potential as a tool for civic empowerment.

One of the most pressing challenges is the digital divide, which refers to the gap between individuals who have access to digital technologies and those who do not. While internet connectivity has expanded globally, disparities persist due to socioeconomic factors, geographical limitations, and infrastructure deficits. People in rural areas, low-income communities, and developing countries often struggle with unreliable internet access, lack of digital literacy, and limited access to devices such as smartphones and computers. This digital inequality creates barriers to civic participation, as those without proper access are unable to engage in online discussions, follow political developments, or participate in digital activism. Additionally, older populations and individuals with disabilities may face further challenges in navigating digital platforms, reducing the inclusiveness of online civic engagement. Without bridging this gap, digital media runs the risk of amplifying existing inequalities rather than democratizing access to information and participation.

Another critical issue is the formation of filter bubbles and echo chambers, where digital media algorithms prioritize content that aligns with users' existing beliefs. Social media platforms use personalized recommendations based on users’ browsing history and interactions, often limiting exposure to diverse viewpoints. As a result, individuals may find themselves in isolated information silos where they only encounter perspectives that reinforce their preconceptions, rather than being exposed to a broad range of political and social perspectives. This phenomenon hinders constructive dialogue and reduces the potential for individuals to develop a well-rounded understanding of civic issues. Moreover, echo chambers deepen political and ideological divides, making it more difficult for different groups to find common ground. When people are only exposed to content that aligns with their views, misinformation and partisan narratives become more deeply ingrained, further complicating efforts to foster informed civic participation(De Zúñiga & Chen, 2019; Ecker et al., 2024).

Compounding these challenges is the widespread proliferation of misinformation and fake news on digital platforms, which can distort public discourse and undermine trust in democratic institutions. False or misleading content spreads rapidly through social media, often fueled by clickbait headlines, conspiracy theories, and politically motivated disinformation campaigns. Misinformation can manipulate public opinion, influence electoral outcomes, and create confusion around important civic issues. The difficulty in distinguishing between credible sources and deceptive content exacerbates this problem, as individuals may unknowingly share or believe false information. Even efforts to debunk misinformation face obstacles, as psychological biases can make people resistant to correcting their beliefs. The presence of misleading content erodes trust in media institutions, creating an environment where people may question all sources of information, thereby weakening the foundation of informed civic engagement(De Zúñiga & Chen, 2019).

Additionally, online harassment, toxicity, and hate speech present significant barriers to productive civic discourse on digital media platforms. While the internet offers a space for individuals to express their opinions, it also exposes them to hostility, cyberbullying, and targeted attacks. Women, minorities, and marginalized groups often face disproportionate levels of online abuse, discouraging them from participating in political discussions or expressing their views openly. The prevalence of online toxicity creates a hostile environment where meaningful conversations are overshadowed by aggression and intimidation. This not only suppresses civic engagement but also discourages individuals from taking part in activism, advocacy, or policy discussions. Despite platform policies aimed at curbing hate speech, enforcement remains inconsistent, and harmful content continues to circulate. If digital spaces remain hostile, individuals may withdraw from civic discussions altogether, limiting the potential of digital media as a democratic tool.

Addressing these challenges requires a collective effort from technology companies, policymakers, civil society organizations, and users. Bridging the digital divide through expanded internet access, improving digital literacy, reforming content algorithms to promote diverse perspectives, implementing stronger misinformation countermeasures, and enforcing anti-harassment policies can help create a healthier digital environment for civic engagement. While digital media holds immense potential for fostering political participation and social activism, overcoming these challenges is essential to ensuring that it remains an inclusive, trustworthy, and empowering space for all citizens.

ETHICAL IMPERATIVES

As digital media continues to shape civic engagement, ethical considerations are paramount in ensuring an inclusive, transparent, and responsible political discourse. The increasing reliance on digital platforms for political discussions, activism, and participation demands adherence to fundamental ethical principles, including inclusivity, accuracy, respect, privacy, and accountability(De Zúñiga & Chen, 2019). By addressing these concerns, digital media can serve as a powerful tool for democratic engagement while mitigating potential risks such as misinformation, polarization, and online toxicity.

One of the most significant ethical imperatives in digital civic engagement is inclusivity. Digital platforms should bridge the digital divide by ensuring that individuals from all socioeconomic, geographic, and demographic backgrounds have access to political discourse and civic participation. This requires addressing disparities in digital literacy, internet access, and technological resources. Marginalized communities, particularly those with limited access to technology, should not be excluded from discussions that impact their lives. Policymakers and platform developers must work toward equitable access, fostering a more representative and diverse political engagement landscape.

Another crucial ethical consideration is accuracy and transparency in the dissemination of political information. The rapid spread of misinformation and disinformation on digital platforms has serious consequences for public perception and democratic decision-making. To counteract this, individuals, organizations, and media outlets must prioritize fact-checking, cite credible sources, and disclose any biases that may influence their content. Transparency in digital civic engagement also requires clear communication about algorithms and content moderation policies, ensuring that users understand how information is filtered and presented to them. Without these safeguards, political discourse risks being manipulated by false narratives and hidden agendas(De Zúñiga & Chen, 2019).

Respectful dialogue is fundamental to ethical civic engagement, as digital media can either foster constructive discussions or amplify hostility and division. Online platforms must establish and enforce community standards that prohibit hate speech, harassment, and personal attacks while promoting respectful interactions. Users should be encouraged to engage in thoughtful discussions, actively listen to opposing viewpoints, and cultivate a culture of civil discourse. Digital media's potential to bring people together across ideological divides depends on creating an environment where individuals feel safe expressing their opinions without fear of abuse or intimidation.

Privacy and data protection are also central ethical concerns in digital civic engagement. Individuals must have control over their personal information, and organizations must handle user data securely and transparently. The increasing use of targeted political advertising and data-driven persuasion strategies raises ethical questions about consent, surveillance, and potential manipulation. Platforms must be transparent about how they collect, store, and utilize user data, ensuring that privacy is not compromised for political or commercial gain. Strong data protection measures can build public trust and prevent the misuse of personal information in digital political campaigns.

Finally, accountability is essential in maintaining ethical standards in digital civic engagement. Organizations, political actors, and media platforms should be held responsible for the accuracy of the information they share and the consequences of their actions. This includes swiftly correcting errors, addressing the spread of misinformation, and taking proactive steps to prevent harm caused by false narratives or unethical practices. Digital platforms should also be accountable for the algorithms they employ, ensuring that they do not promote sensationalism, divisiveness, or misleading content in pursuit of engagement metrics(De Zúñiga & Chen, 2019).

By adhering to these ethical imperatives, digital media can effectively promote civic engagement while upholding democratic values. Ensuring inclusivity, accuracy, respectful dialogue, privacy protection, and accountability will help foster a more informed and engaged society. While digital platforms provide unprecedented opportunities for participation and discourse, their potential must be harnessed responsibly to prevent harm and safeguard the integrity of democratic engagement.

DIGITAL MEDIA'S INFLUENCE

Digital media has transformed political discourse by reshaping how information is disseminated, shared, and consumed, making political engagement more accessible and dynamic. It has democratized access to political information, allowing individuals to participate in discussions regardless of their location or socioeconomic status. Social media platforms, online news outlets, and political blogs enable people to express their opinions, engage in debates, and reach a wider audience. Additionally, digital media amplifies the voices of historically marginalized or underrepresented groups, fostering a more inclusive political landscape. Research by Chadwick et al. (2017) highlights how non-elite individuals influence political discourse through social media, impacting public perception and engagement, as seen during the 2015 UK general election debate. Furthermore, digital media facilitates real-time communication, allowing instantaneous updates, live discussions, and immediate reactions to unfolding political events. However, its viral nature also poses challenges, as political messages and news stories spread rapidly, sometimes without verification, leading to misinformation and shaping public opinion in potentially misleading ways. While digital media enhances political participation and engagement, its influence must be navigated responsibly to ensure a well-informed and balanced discourse.

DIGITAL MEDIA IN POLITICAL DISCOURSE

Digital media has profoundly reshaped political discourse, providing both significant benefits and notable challenges. It has revolutionized the way individuals engage with politics, breaking down traditional barriers to participation and fostering a more inclusive and dynamic public sphere. The accessibility and immediacy of digital platforms enable people, especially younger and more technologically adept generations, to voice their opinions, mobilize for causes, and interact directly with political leaders. This transformation has given rise to new forms of activism and engagement, allowing for rapid organization and mobilization around key issues. Additionally, digital media has played a crucial role in democratizing information, ensuring that underrepresented and marginalized voices can participate in political conversations, challenging mainstream narratives, and offering alternative perspectives.

Another key advantage of digital media is its ability to disseminate information at an unprecedented speed. Political events, debates, and policy changes are reported and shared in real-time, enabling individuals to stay informed and respond swiftly. Social media platforms, blogs, and independent news sources provide an extensive range of content, fostering a more pluralistic discourse where multiple viewpoints can coexist. This instant access to information allows for greater political awareness and engagement, as individuals can easily access news, fact-check claims, and critically analyze political developments. Furthermore, digital media provides opportunities for direct interaction between politicians and the public, reducing the reliance on traditional gatekeepers such as mainstream media and enabling more transparent and accountable governance.

Despite these advantages, digital media also presents significant challenges that can undermine the quality of political discourse. One of the most concerning issues is the emergence of filter bubbles and echo chambers, where individuals are exposed primarily to information that aligns with their existing beliefs. Algorithms on social media platforms are designed to personalize content based on user preferences, reinforcing biases and limiting exposure to diverse viewpoints. This selective exposure can contribute to ideological polarization, as individuals become less open to opposing perspectives and more entrenched in their beliefs. As a result, meaningful dialogue and critical thinking can be stifled, reducing the potential for constructive debate.

Misinformation and disinformation further complicate the landscape of digital political discourse. The rapid spread of false or misleading content on social media platforms can distort public perception and fuel division. Fake news, propaganda, and conspiracy theories can be disseminated widely before they are fact-checked or debunked, shaping public opinion in ways that may not align with reality(De Zúñiga & Chen, 2019). The difficulty in distinguishing credible sources from unreliable ones adds another layer of complexity, as misinformation can manipulate narratives, influence elections, and erode trust in democratic institutions.

Moreover, the anonymity of digital platforms often contributes to a toxic online environment where incivility thrives. The lack of face-to-face interaction can embolden individuals to engage in harassment, personal attacks, and hate speech, creating an atmosphere of hostility that discourages meaningful political discussions. Politicians, journalists, activists, and ordinary citizens alike have faced online threats and abuse, which can silence critical voices and undermine democratic engagement. The spread of divisive and inflammatory rhetoric can further escalate tensions, making it more challenging to foster productive political conversations.

While digital media provides unparalleled opportunities for political engagement and discourse, its risks must be carefully managed. Addressing issues such as misinformation, filter bubbles, and online incivility requires collective efforts from governments, technology companies, media organizations, and the public. Promoting media literacy, encouraging critical thinking, and implementing policies to curb the spread of false information can help create a more informed and balanced digital space(De Zúñiga & Chen, 2019). At the same time, fostering respectful and open discussions can contribute to a healthier democratic environment where diverse voices are heard and valued. In navigating the complexities of digital media, it is essential to strike a balance between harnessing its potential for political empowerment and mitigating its challenges to ensure a more constructive and inclusive discourse.

ETHICAL RESPONSIBILITIES

Ethical considerations play a fundamental role in ensuring that digital media serves as a platform for constructive, inclusive, and responsible political discourse. As digital platforms increasingly shape public opinion, electoral processes, and policy discussions, upholding ethical principles is critical to maintaining democratic values and fostering informed civic engagement. One of the most crucial ethical imperatives is the commitment to accuracy and truthfulness in political discourse. Misinformation, disinformation, and misleading narratives can easily spread across digital platforms, distorting public perception and eroding trust in democratic institutions. To mitigate this, individuals, media organizations, and policymakers must prioritize fact-checking, verify claims before dissemination, and promptly correct inaccuracies when they arise. Digital media users should also develop critical thinking skills to assess sources, detect bias, and distinguish between credible information and manipulative content (Allcott & Gentzkow, 2017; Lazer et al., 2018).

Transparency and accountability are equally essential in digital political discourse. The sources of information, including political advertisements, sponsored content, and opinion pieces, should be clearly disclosed to avoid misleading the public. Users must be aware of potential biases, conflicts of interest, and the motivations behind the content they consume. Additionally, individuals and organizations that contribute to political discourse should take responsibility for their statements, ensuring that their messages do not incite violence, propagate falsehoods, or manipulate public opinion for unethical gains. Governments, regulatory bodies, and digital platforms must also implement and enforce policies that promote transparency, preventing the misuse of digital media for propaganda or deceptive practices. Holding individuals and institutions accountable for their actions in the digital space helps build a more trustworthy and reliable political discourse (Matheus et al., 2023).

Another crucial ethical principle is fostering respectful dialogue and inclusivity in political discussions on digital media. While these platforms provide opportunities for open debate, they also create environments where polarization, hostility, and hate speech can thrive. Ethical engagement requires users to maintain civil discourse, respect diverse perspectives, and engage in conversations that contribute to constructive political debate rather than deepen divisions. Digital platforms must actively combat discrimination, hate speech, and harassment, ensuring that all voices—especially those from marginalized communities—can participate in the political conversation without fear of personal attacks. Promoting respectful engagement also involves challenging harmful narratives and misinformation without resorting to personal insults or aggressive rhetoric (Pukallus & Arthur, 2024).

Privacy and data protection are fundamental ethical concerns in digital political discourse, particularly in an era where personal data is extensively collected, analyzed, and sometimes exploited for political purposes. Individuals should have control over how their personal information is collected, stored, and used, and they should be aware of how digital platforms track their online behavior to influence political messaging. The unethical use of data analytics, such as microtargeting voters based on their psychological profiles, can manipulate political decisions and undermine the democratic process. To uphold ethical standards, policymakers must enforce robust data protection laws, while technology companies should implement clearer data usage policies and offer users more control over their personal information (Van Gils et al., 2024).

Lastly, algorithmic transparency and fairness are crucial in shaping political discourse in the digital age. Social media and search engine algorithms influence what information users see, potentially creating echo chambers that reinforce biases and limit exposure to diverse perspectives. If these algorithms prioritize sensationalized or divisive content for engagement-driven profit motives, they can distort political discourse and manipulate public perception. Ensuring that algorithms are transparent, impartial, and designed to promote factual and balanced information is essential to fostering ethical digital engagement (Cheong, 2024). By adhering to these ethical considerations—accuracy, transparency, respectful discourse, privacy protection, and algorithmic fairness—digital media can play a constructive role in political discussions, enabling informed civic participation while upholding democratic principles.

SHAPING PUBLIC OPINION IN THE DIGITAL AGE

Digital media significantly influences public opinion by transforming how individuals' access, interpret, and express views on social, political, and economic issues. Unlike traditional media, digital platforms enable real-time information dissemination, allowing people to fact-check claims, access diverse perspectives, and stay informed about global events. This accessibility decentralizes information control, empowering independent journalists, grassroots movements, and marginalized communities to share alternative narratives often overlooked by mainstream media (Kaplan & Haenlein, 2009).

A key impact of digital media is its ability to amplify voices. Social media, blogs, and online forums provide individuals with platforms to shape narratives, mobilize movements, and engage in public discourse. Hashtag activism, viral campaigns, and online petitions demonstrate its role in influencing collective opinions and driving real-world action. However, this democratization also poses challenges, including misinformation and the spread of opinion-based narratives that lack factual accuracy. The ease of publishing online content allows false information to circulate widely, affecting public perception and eroding trust in credible sources (De Zúñiga & Chen, 2019).

Algorithm-driven content personalization further affects opinion formation. Social media platforms and search engines curate content based on user preferences, creating filter bubbles that reinforce existing beliefs and limit exposure to diverse perspectives. This echo chamber effect hinders critical discussions and promotes polarization. Additionally, engagement-driven content prioritizes sensationalized or emotionally charged material, shaping public attitudes through curated and sometimes misleading narratives (Diakopoulos & Koliska, 2016).

The virality of digital media accelerates opinion formation. Content spreads rapidly, influencing sentiment within minutes. While this can drive social movements and policy discussions, it also enables manipulation by politically or commercially motivated actors. Recognizing these dynamics is essential for fostering media literacy, critical thinking, and responsible digital engagement. By addressing misinformation, polarization, and algorithmic bias, society can ensure digital media serves as a tool for informed, democratic discourse.

EXAMPLES OF HOW DIGITAL MEDIA HAS INFLUENCED PUBLIC OPINION.

MeToo Movement

The #MeToo movement acquired traction through digital media platforms, which allowed individuals to share their experiences and raise awareness of sexual harassment and assault. Social media provided a forum for survivors to share their experiences, influencing public opinion regarding the prevalence and impact of these issues(Romero, 2021; Shukla et al., 2020).

Political Campaigns

Digital media has become an indispensable instrument for political campaigns to influence public opinion. Candidates and political parties use social media platforms to disseminate messages, interact with electors, and shape public perception. Online advertising, viral content, and social media campaigns can significantly impact public opinion during elections(Rajpopat, 2024).

Climate Change Activism

Digital media have played a crucial role in promoting public awareness and mobilising public opinion regarding climate change. Online campaigns, influential social media personalities, and the dissemination of environmental content have contributed to the pervasive recognition of the issue and the rise in public support for climate action (Andersen, 2024).

NAVIGATING THE CHALLENGES OF DIGITAL MEDIA'S INFLUENCE ON PUBLIC OPINION

While digital media has revolutionized the way public opinion is shaped, several challenges hinder its ability to provide a balanced and accurate representation of information. One of the most pressing issues is the prevalence of filter bubbles and echo chambers, which arise from the personalized content algorithms used by digital media platforms. These algorithms analyze user behavior, preferences, and interactions to curate content that aligns with an individual's existing beliefs and interests. While this personalization enhances user engagement, it also creates informational silos, where individuals are primarily exposed to viewpoints that reinforce their opinions rather than challenge them. As a result, people may develop a skewed understanding of issues, leading to polarization and resistance to alternative perspectives. The lack of diverse exposure prevents constructive dialogue, making it difficult for societies to foster well-rounded, informed public opinions.

Another significant challenge is the proliferation of misinformation and disinformation on digital media platforms. Unlike traditional journalism, which adheres to editorial standards and fact-checking processes, digital media enables the rapid spread of unverified information. Misinformation, which refers to false information shared unintentionally, and disinformation, which is deliberately misleading content, can distort public perception of critical issues. Viral hoaxes, deepfake videos, and misleading headlines can manipulate audiences, influencing public discourse and shaping opinions based on falsehoods. The speed at which misinformation spreads, combined with confirmation bias—where individuals believe information that aligns with their preexisting views—exacerbates the problem, making it difficult to correct false narratives once they take hold. This erosion of trust in information sources can have severe consequences for democracy, decision-making, and public trust in institutions (KaabOmeir et al., 2024).

A further complication arises from algorithmic biases embedded in digital media platforms. While algorithms are designed to optimize user experience by recommending relevant content, they often reflect and amplify societal biases present in the data they process. These biases can influence the visibility and accessibility of certain perspectives, favoring dominant narratives while marginalizing alternative voices. For instance, political content may be prioritized based on engagement metrics rather than factual accuracy, meaning emotionally charged or sensationalized content often receives more visibility than nuanced, well-researched discussions. Algorithmic biases can also affect representation by reinforcing stereotypes, silencing dissenting voices, and shaping public opinion through a distorted lens. Without transparency in how these algorithms function, users may unknowingly be subjected to an environment that limits their exposure to a comprehensive range of viewpoints (Cakmak et al., 2024).

Compounding these issues is the lack of media literacy among digital media consumers. In the digital age, the ability to critically assess, interpret, and verify information is essential. However, many individuals lack the skills necessary to differentiate between credible sources and misleading content. Without proper media literacy education, users are more susceptible to manipulation, propaganda, and the uncritical acceptance of information that aligns with their biases. This lack of critical engagement contributes to the spread of misinformation, polarization, and uninformed decision-making. Addressing these challenges requires a multi-faceted approach, including promoting media literacy programs, advocating for algorithmic transparency, and fostering a diverse and inclusive digital media ecosystem that encourages exposure to multiple perspectives (Diakopoulos & Koliska, 2016; Pklein, 2022). By implementing these measures, societies can mitigate the negative effects of digital media on public opinion and ensure a more informed and engaged citizenry.

RECOMMENDATIONS AND SOLUTIONS

To maximize the benefits of digital media while mitigating its challenges, a multi-stakeholder approach involving policymakers, technology companies, educators, and civil society is necessary. Digital media has significantly reshaped public opinion, but its challenges, such as misinformation, filter bubbles, algorithmic bias, and online harassment, demand comprehensive solutions. Combating misinformation and disinformation requires strengthening independent fact-checking initiatives that verify news accuracy and provide real-time corrections to misleading content. Digital platforms must enhance algorithmic detection mechanisms using artificial intelligence to identify and flag false or manipulated information before it spreads widely. Promoting responsible journalism is also crucial, as media organizations must commit to transparent reporting, while digital platforms should collaborate with credible sources to prioritize verified news. Additionally, empowering users with robust reporting tools ensures a community-driven approach to content moderation, making it easier for individuals to flag misinformation and contribute to maintaining the integrity of digital discourse. Addressing filter bubbles and echo chambers is another critical challenge that requires improving algorithmic transparency so that users understand how content is curated and have greater control over their digital experiences. Encouraging exposure to diverse perspectives can be facilitated by digital platforms introducing features such as "Opposing Viewpoints" sections, fact-checked alternative narratives, and diversified newsfeeds that counteract ideological silos. Furthermore, integrating digital literacy programs in schools and workplaces can help individuals develop critical thinking skills, enabling them to assess online information accurately, identify bias, and recognize manipulation tactics used to sway public opinion. Reducing online harassment and promoting respectful dialogue necessitate enforcing stricter content moderation policies to combat hate speech, cyberbullying, and disinformation campaigns. Platforms should implement AI-powered moderation systems to detect and mitigate toxic interactions, while user-controlled comment filtering options can help individuals personalize their digital experiences and minimize exposure to harmful discussions. Algorithmic bias remains a pressing concern, as content curation systems may disproportionately promote specific narratives or suppress minority viewpoints. To address this, technology companies must develop fair and inclusive AI models that undergo regular bias audits, ensuring that digital media platforms remain equitable and diverse. Additionally, governments should increase regulatory oversight to enforce ethical AI guidelines and prevent manipulative content dissemination. Encouraging meaningful civic engagement through digital media involves leveraging online platforms to promote offline participation, such as voter mobilization efforts, petitions, and grassroots organizing. Community-based digital forums should be strengthened to provide safe and inclusive spaces where individuals can engage in thoughtful discourse without fear of online attacks. Emerging technologies, including artificial intelligence, virtual reality, and blockchain, offer innovative solutions for enhancing transparency in political discourse and securing electoral integrity. By implementing these recommendations, digital media can evolve into a tool that fosters informed discourse, promotes inclusivity, and strengthens democratic participation. A collaborative effort among key stakeholders—policymakers, technology companies, educators, and civil society—is essential to ensure that digital platforms contribute positively to society, rather than exacerbating misinformation, bias, and division. Only through collective responsibility and strategic interventions can digital media fulfil its potential as a force for good in shaping public opinion and democratic engagement.

FUTURE IMPLICATIONS

As digital media continues to evolve, future research must explore its influence on public opinion and political discourse to ensure that these platforms contribute positively to civic engagement and democracy. One crucial area of study involves understanding the dynamics of online echo chambers and filter bubbles and their impact on shaping public perception. Research should focus on how these digital silos emerge, the extent to which they reinforce biases, and their effects on political polarization and social fragmentation. By analyzing user behaviors, algorithmic curation, and the role of social networking in reinforcing ideological divides, scholars can identify strategies to foster more open and balanced online discussions.

Another critical research avenue is examining the effectiveness of strategies designed to combat misinformation and enhance media literacy. As misinformation and disinformation proliferate across digital platforms, there is a pressing need to evaluate the success of fact-checking initiatives, algorithmic interventions, and educational campaigns in mitigating their spread. Research should assess the psychological and cognitive factors influencing people's susceptibility to misinformation, as well as the role of digital media companies in curbing false narratives. Additionally, studying how different demographics respond to misinformation—and the effectiveness of interventions targeted at diverse populations—can help tailor more impactful solutions to ensure that users develop critical thinking skills and digital literacy competencies.

The ethical implications of algorithmic biases in digital media platforms also warrant deeper investigation. Since algorithms determine what content users see, their biases can shape political narratives and limit exposure to diverse perspectives. Research should explore how these biases emerge, how they impact marginalized voices, and what regulatory or technological measures can be implemented to promote fair and balanced content distribution. Further, transparency in algorithmic decision-making processes is crucial for ensuring public trust in digital platforms. Understanding how algorithms influence political discourse, voting behavior, and civic participation can inform the development of ethical guidelines that balance engagement-driven content delivery with responsible information dissemination.

Additionally, investigating the influence of emerging technologies such as artificial intelligence (AI) and virtual reality (VR) on public opinion and political discourse is an essential area for future study. AI-driven content generation, deepfake technology, and personalized digital assistants all have the potential to transform how individuals consume political information. Research should explore both the risks and benefits of these technologies, including their role in spreading misinformation, manipulating public perception, and shaping political ideologies. Similarly, VR-based political engagement—such as virtual town halls and immersive journalism—could revolutionize how people interact with political content, necessitating an analysis of its long-term implications.

Finally, assessing the long-term effects of digital media on political participation, civic engagement, and democratic processes is critical. While digital media has enabled new forms of activism and engagement, questions remain about whether these interactions translate into meaningful offline participation. Future studies should investigate how online activism compares to traditional forms of civic involvement, the sustainability of digital movements, and whether digital engagement strengthens or weakens democratic institutions over time.

In practice, policymakers, digital media platforms, and civil society organizations must collaborate to develop guidelines, policies, and initiatives that foster the responsible use of digital media. Policymakers should implement regulations that promote transparency, protect users from manipulation, and hold platforms accountable for misinformation and harmful content. Digital media companies must invest in ethical algorithm development, improve content moderation, and support initiatives that encourage informed public discourse. Civil society organizations should play an active role in media literacy education, advocating for inclusivity and ensuring that marginalized voices are heard in digital spaces.

By addressing these implications, societies can harness the power of digital media to create a more informed, engaged, and participatory public sphere. Through interdisciplinary research and collaborative policy efforts, digital media can become a tool that strengthens democracy rather than undermines it, ensuring a future where political discourse is both transparent and inclusive.

DISCUSSION

The role of digital media in shaping public opinion and political discourse is multifaceted, offering both opportunities and challenges. As explored in this chapter, digital media has transformed how information is disseminated, accessed, and shared, influencing public perception and civic engagement. However, alongside its benefits, digital media also presents challenges, such as the spread of misinformation and the erosion of traditional media's role in political discourse.

A key aspect of digital media's influence is its role in political campaigns. By providing a direct channel of communication between candidates and voters, social media platforms have become indispensable tools for mobilizing support and shaping public narratives. Numerous examples highlight how digital campaigns have successfully influenced political discourse, demonstrating the effectiveness of these platforms in reaching and engaging the public.

Despite its advantages, digital media also facilitates the spread of misinformation, which can significantly impact public opinion. Theories such as agenda-setting and the networked public sphere provide insights into how information—both accurate and misleading—spreads and influences political discussions. Efforts to counteract misinformation, including fact-checking initiatives and digital literacy programs, are essential to maintaining the integrity of political discourse.

Another critical dimension is the potential of digital media to enhance civic engagement. By enabling participation, mobilization, and the amplification of diverse voices, digital media platforms can foster a more engaged citizenry. However, challenges remain, including ethical concerns surrounding digital activism and the risk of echo chambers reinforcing existing biases rather than encouraging informed debate.

Research findings, such as those by David (2021), illustrate how digital media usage—particularly on platforms like Facebook—can shape individual support for social movements. His study on the Israeli social movement highlights how perceptions of public support and personal involvement are key predictors of activism. However, it also reveals that legacy media has a diminished role in influencing these perceptions, suggesting that digital platforms now play a more significant role in shaping collective action.

Burgess & Matamoros-Fernández (2016) stress the need for empirical methods to examine digital controversies, highlighting sociocultural dynamics and the role of media objects in shaping debates. They advocate for issue-mapping techniques that include minority voices while acknowledging challenges in analyzing multimedia content across platforms. These discussions underscore the complexity of digital media's influence on public opinion. While it fosters engagement, responsible use, media literacy, and critical evaluation are vital to maintaining an informed public sphere. Combating misinformation, polarization, and manipulation is essential to ensuring digital media supports democratic discourse rather than deepening divisions.

CONCLUSION

Digital media has become a vital tool for political communication, shaping public opinion and engagement. While it enhances political campaigns, the rise of misinformation underscores the need for ethical use. Balancing its benefits and risks is crucial to fostering democratic dialogue. This study highlights the significance of responsible digital media practices in political discourse. Future research should explore strategies to optimize its impact while mitigating challenges. Ensuring accurate information dissemination and constructive engagement is essential for a thriving democracy. A thoughtful approach to digital media use can strengthen public trust and participation in political processes.

REFERENCES

Abubakari, M. R., DiNicola, K., & Lee, W. (2020). Empirical insights on technology use for navigating human services. *Journal of Technology in Human Services, 38*(4), 331–351. DOI: 10.1080/15228835.2020.1789025

Alalawi, S., Baalfaqih, S., Almeqbaali, M., & Masud, M. M. (2023). Social media misinformation propagation and detection. In *Proceedings of the 15th International Conference on Innovations in Information Technology (IIT 2023)* (pp. 240–245). IEEE. https://doi.org/DOI: 10.1109/IIT59782.2023.10366483

Allcott, H., & Gentzkow, M. (2017). Social media and fake news in the 2016 election. *The Journal of Economic Perspectives, 31*(2), 211–236. DOI: 10.1257/jep.31.2.211

Andersen, C. H. (2024, February 1). Meet the social media activists changing the narrative on climate change. *Reader's Digest.* https://www.rd.com/article/social-media-activists-climate-change/

Banciu, D., & Cirnu, C. E. (2022). AI Ethics and Data Privacy compliance. *2022 14th International Conference on Electronics, Computers and Artificial Intelligence (ECAI)*, 1–5. https://doi.org/DOI: 10.1109/ecai54874.2022.9847510

Bernacki, M. L., Greene, M. J., & Lobczowski, N. G. (2021). A Systematic Review of Research on Personalized Learning: Personalized by Whom, to What, How, and for What Purpose(s)? *Educational Psychology Review, 33*(4), 1675–1715. DOI: 10.1007/s10648-021-09615-8

Burgess, J., & Matamoros-Fernández, A. (2016). Mapping sociocultural controversies across digital media platforms: One week of# gamergate on Twitter, YouTube, and Tumblr. *Communication Research and Practice, 2*(1), 79–96. DOI: 10.1080/22041451.2016.1155338

Cakmak, M. C., Agarwal, N., & Oni, R. (2024). The bias beneath: Analyzing drift in YouTube's algorithmic recommendations. *Social Network Analysis and Mining, 14*(1). Advance online publication. DOI: 10.1007/s13278-024-01343-5

Chadwick, A. (2013). *The hybrid media system: Politics and power*. Oxford University Press., DOI: 10.1093/acprof:oso/9780199661992.001.0001

Chadwick, A., O'Loughlin, B., & Vaccari, C. (2017). Why people dual screen political debates and why it matters for democratic engagement. *Journal of Broadcasting & Electronic Media, 61*(2), 220–239. DOI: 10.1080/08838151.2017.1309415

Chau, X. T. D., Nguyen, T. T., Jo, J., & Nguyen, Q. V. H. (2022). Are rumors always false? Understanding rumors across domains, queries, and ratings. In Vol. 13087, pp. 174–189). Lecture Notes in Computer Science. Springer., DOI: 10.1007/978-3-030-95405-5_13

Cheong, B. C. (2024). Transparency and accountability in AI systems: Safeguarding wellbeing in the age of algorithmic decision-making. *Frontiers in Human Dynamics, 6*. Advance online publication. DOI: 10.3389/fhumd.2024.1421273

Copeland, L., & Bimber, B. (2014). Variation in the relationship between digital media use and political participation in U.S. elections over time, 1996–2012: Does Obama's reelection change the picture? *Journal of Information Technology & Politics, 12*(1), 74–87. DOI: 10.1080/19331681.2014.975391

David, Y. (2021). Public opinion, media and activism: The differentiating role of media use and perceptions of public opinion on political behaviour. *Social Movement Studies, 21*(3), 334–354. DOI: 10.1080/14742837.2021.1875321

Davis, S. D., Chan, J. C., & Wilford, M. M. (2017). The dark side of interpolated testing: Frequent switching between retrieval and encoding impairs new learning. *Journal of Applied Research in Memory and Cognition, 6*(4), 434–441. DOI: 10.1016/j.jarmac.2017.07.002

De Zúñiga, H. G., & Chen, H. (2019). Digital Media and Politics: Effects of the great Information and Communication Divides. *Journal of Broadcasting & Electronic Media, 63*(3), 365–373. DOI: 10.1080/08838151.2019.1662019

Diakopoulos, N., & Koliska, M. (2016). Algorithmic transparency in the news media. *Digital Journalism (Abingdon, England)*, *5*(7), 809–828. DOI: 10.1080/21670811.2016.1208053

Dimitrova, D. V., & Matthes, J. (2018). Social media in political campaigning around the world: Theoretical and methodological challenges. *Journalism & Mass Communication Quarterly*, *95*(2), 333–342. DOI: 10.1177/1077699018770437

Du Chau, X. T., Nguyen, T. T., Jo, J., & Nguyen, Q. V. H. (2022). Are rumors always false?: Understanding rumors across domains, queries, and ratings. In *Lecture notes in computer science* (pp. 174–189). https://doi.org/DOI: 10.1007/978-3-030-95405-5_13

Ecker, U., Roozenbeek, J., Van Der Linden, S., Tay, L. Q., Cook, J., Oreskes, N., & Lewandowsky, S. (2024). Misinformation poses a bigger threat to democracy than you might think. *Nature*, *630*(8015), 29–32. DOI: 10.1038/d41586-024-01587-3 PMID: 38840026

Getachew, A., & Beshah, T. (2019). The role of social media in citizen's political participation. In *IFIP advances in information and communication technology* (pp. 487–496). https://doi.org/DOI: 10.1007/978-3-030-20671-0_33

Hindawy, E.-S. F. M. (2021). Modern administrative methods to combat electronic rumors. *Journal of Legal* [Allied Business Academies.]. *Ethical and Regulatory Issues*, *24*(1), 1–11.

Jost, P., Kruschinski, S., Sülflow, M., Haßler, J., & Maurer, M. (2023). Invisible transparency: How different types of ad disclaimers on Facebook affect whether and how digital political advertising is perceived. *Policy and Internet*, *15*(2), 204–222. DOI: 10.1002/poi3.333

KaabOmeir, F., Khademizadeh, S., Seifadini, R., Balani, S. O., & Khazaneha, M. (2024). Overview of Misinformation and Disinformation Research from 1971 to 2022. *Journal of Scientometric Research*, *13*(2), 430–447. DOI: 10.5530/jscires.13.2.34

Kaplan, A. M., & Haenlein, M. (2009). Users of the world, unite! The challenges and opportunities of Social Media. *Business Horizons*, *53*(1), 59–68. DOI: 10.1016/j.bushor.2009.09.003

Kaur, M., & Verma, R. (2018). Demographics, social media usage, and political engagement in Punjab. *Indian Journal of Marketing*, *48*(11), 43. DOI: 10.17010/ijom/2018/v48/i11/137984

Kaur, S., & Kaur, M. (2013). Impact of social media on politics. *Gian Jyoti E-Journal, 3*(4), 1-XX. ISSN 2250-348X

Kelman, H. C. (1958). Compliance, identification, and internalization three processes of attitude change. *The Journal of Conflict Resolution*, *2*(1), 51–60. DOI: 10.1177/002200275800200106

Kgatla, L. (2024). Exploring ethical complexities in social media data mining: Questions from a researcher's journey investigating depression. *Educational Research for Social Change*, *13*(2), 65–79. DOI: 10.5281/zenodo.14031438

Komodromos, M. (2015). Social media and its role for Cypriot members of parliament in times of crisis. *The Cyprus Review*, *27*(1), 217–232.

Kumar, P. (2023, November 5). The Power of Media in Shaping Public Opinion • Journalism University. *Journalism & Mass Communication Hub*. https://journalism.university/media-and-society/power-media-shaping-public-opinion

Kwon, K. H., Bang, C. C., Egnoto, M. J., & Rao, H. R. (2016). Social media rumors as improvised public opinion: Semantic network analyses of twitter discourses during Korean saber rattling 2013. *Asian Journal of Communication*, *26*(3), 201–222. DOI: 10.1080/01292986.2015.1130157

Lazer, D. M. J., Baum, M. A., Benkler, Y., Berinsky, A. J., Greenhill, K. M., Menczer, F., Metzger, M. J., Nyhan, B., Pennycook, G., Rothschild, D., Schudson, M., Sloman, S. A., Sunstein, C. R., Thorson, E. A., Watts, D. J., & Zittrain, J. L. (2018). The science of fake news. *Science*, *359*(6380), 1094–1096. DOI: 10.1126/science.aao2998 PMID: 29590025

López-López, P. C., Barredo-Ibáñez, D., & Jaráiz-Gulías, E. (2023). Research on digital political communication: Electoral campaigns, disinformation, and artificial intelligence. *Societies (Basel, Switzerland)*, *13*(5), 126. DOI: 10.3390/soc13050126

Matheus, R., Faber, R., Ismagilova, E., & Janssen, M. (2023). Digital transparency and the usefulness for open government. *International Journal of Information Management*, *73*, 102690. DOI: 10.1016/j.ijinfomgt.2023.102690

Nickerson, R. S. (1998). Confirmation bias: A ubiquitous phenomenon in many guises. *Review of General Psychology*, *2*(2), 175–220. DOI: 10.1037/1089-2680.2.2.175

Oparaugo, B. (2021). Media and Politics: Political Communication in the Digital Age. SSRN *Electronic Journal*. https://doi.org/DOI: 10.2139/ssrn.3780554

Oser, J., & Boulianne, S. (2020). Reinforcement Effects between Digital Media Use and Political Participation: A Meta-Analysis of Repeated-Wave Panel Data. *Public Opinion Quarterly*, *84*(S1), 355–365. DOI: 10.1093/poq/nfaa017

Patalauskaitė, E. (2024). Ethical aspects of content creation. *Filosofija Sociologija*, *35*(3). Advance online publication. DOI: 10.6001/filsoc.2024.35.3.14

Pklein. (2022, April 5). *Transparency: The first step to fixing social Media - MIT Initiative on the Digital Economy*. MIT Initiative on the Digital Economy. https://ide.mit.edu/insights/transparency-the-first-step-to-fixing-social-media/

Pukallus, S., & Arthur, C. (2024). Combating hate speech on social media: Applying targeted regulation, developing Civil-Communicative skills and utilising local Evidence-Based Anti-Hate speech interventions. *Journalism and Media*, *5*(2), 467–484. DOI: 10.3390/journalmedia5020031

Rajpopat, V. (2024). The impact of digital media on political campaigns and voter behavior. In *Unified Visions: Collaborative Paths in Multidisciplinary Research-2*. https://doi.org/DOI: 10.25215/8198189815.17

Ramesh, G. S., Supriya, K. H. S., Akash, P., Rukmananda Reddy, A., Tejaswini, V., & Dharmireddi, S. (2024). Fake news detection on social media using a stacking model. In *Proceedings of the 5th International Conference on Computer and Communication Technologies (IC3T 2023)* (pp. 391–402). Springer. https://doi.org/DOI: 10.1007/978-981-99-9707-7_37

Ray, A., & George, J. F. (2019). Online disinformation and the psychological bases of prejudice and political conservatism. In Bui, T. X. (Ed.), *Proceedings of the 52nd Annual Hawaii International Conference on System Sciences (HICSS 2019)* (pp. 2742–2752). IEEE Computer Society.

Ribeiro, S. G. (2024). Freedom of speech and the informer's duty to the voter: Ethics in electoral communication in Brazil. *Cadernos de Dereito Actual, 26*, 120–151. DOI: 10.5281/zenodo.14172082

Rodríguez-Martínez, G., & Lozano, C. a. A. (2021). Uso de internet y redes sociales en el marco de la contingencia Covid-19 en Colombia. *Texto Livre Linguagem E Tecnologia, 15*, e34828. https://doi.org/DOI: 10.35699/1983-3652.2022.34828

Romero, D. F. (2021). Digital footprints of #MeToo. www.academia.edu. https://www.academia.edu/114707913/Digital_footprints_of_MeToo

Sapar, N., & Hussin, A. R. C. (2024). Role of attitude, norm and behaviour control among young voters in social media toward political engagement. In *Lecture notes on data engineering and communications technologies* (pp. 150–159). https://doi.org/DOI: 10.1007/978-3-031-59711-4_14

Sheth, S. (2019). Super PACs, personal data, and campaign finance loopholes. *Virginia Law Review, 105*(3), 655–702.

Shukla, S., Singh, P. P., & Garima. (2020). #MeToo movement: Influence of social media engagement on intention to control sexual harassment against women. *Journal of Content, Community & Communication, 12*(6)

Theocharis, Y., Barberá, P., Fazekas, Z., Popa, S. A., & Parnet, O. (2016). A bad workman blames his tweets: The consequences of citizens' uncivil Twitter use when interacting with party candidates. *Journal of Communication, 66*(6), 1007–1031. DOI: 10.1111/jcom.12259

Tirado-Morueta, R., Mendoza-Zambrano, D., Marín-Gutiérrez, I., & Mendoza-Zambrano, M. (2017). The relativity of sociodemographic determinism on the digital divide in high school students in Ecuador. *International Journal of Communication, 11*, 1528–1551.

Trindade, L. V. P. (2023). Disability is no laughing matter. A critical discourse analysis of disability mockery on social media in Brazil. *Canadian Journal of Latin American and Caribbean Studies / Revue Canadienne Des Études Latino-américaines Et Caraïbes, 49*(1), 22–36. https://doi.org/DOI: 10.1080/08263663.2023.2271306

Van Gils, F., Müller, W., & Prüfer, J. (2024). Microtargeting, voters' unawareness, and democracy. *Journal of Law Economics and Organization*. Advance online publication. DOI: 10.1093/jleo/ewae002

Wagner, M. C., & Boczkowski, P. J. (2019). The reception of fake news: The interpretations and practices that shape the consumption of perceived misinformation. *Digital Journalism (Abingdon, England), 7*(7), 870–885. DOI: 10.1080/21670811.2019.1653208

Wicker, A. W., & Doyle, J. (2012). Leveraging multiple mechanisms for information propagation. In *Lecture notes in computer science* (pp. 1–2). https://doi.org/DOI: 10.1007/978-3-642-27216-5_1

Willis, S., & Tranter, B. (2006). Beyond the 'digital divide.'. *Journal of Sociology (Melbourne, Vic.), 42*(1), 43–59. DOI: 10.1177/1440783306061352

Wood, A. K., & Ravel, A. M. (2018). Fool me once: Regulating "fake news" and other online advertising. *Southern California Law Review, 91*(6), 1223–1278.

Chapter 2
Changing Dynamics of Political Communication in the Digital Era:
A Case Study on India

Romica Bhat
Amity University, Rajarhat, India

Sayantan Ghosh
Amity University, Rajarhat, India

ABSTRACT

This case study examines how political communication in India has evolved with the rise of digital technology, analyzing its impact on political campaigns, voter engagement, and public discourse. The study delves into the significance of social media platforms, digital messaging apps, and online news outlets in shaping the political landscape in the world's largest democracy.In recent years, India has witnessed a significant surge in internet and smartphone usage, leading to a transformation in the way political parties and leaders interact with their constituents. This study investigates the various tools and strategies employed by political actors to connect with voters and disseminate their messages in a digital-first environment.

DOI: 10.4018/978-1-6684-9356-4.ch002

INTRODUCTION

Communication has always been an integral function of acquiring power and has been used to bring about changes in the social and political world around us (Das & Alam, 2019). According to the International Encyclopaedia of the Social and Behavioural Sciences, "Political communication is an interactive process in the transmission of information among politicians, the news media, and the public." "The process operates downward from governing institutions towards citizens, horizontally in linkages among political actors, and also upward from public opinion towards authorities" (Norris, 2000). In any democracy, political communication is a tool to facilitate change. Not only this, but for any deception and manipulation, political communication has taken centre stage and has been part and parcel of democratic politics (Bakir & McStay, 2018).

Since its inception, the media has played a significant role in facilitating political communication in order to reach a wider audience. Traditionally, in political communication, newspapers and radio played key roles (Chadwick, 2017). But with time, the situation changed. Across the world, with the rise of digital media, the fundamental basis of political communication has become digitally dependent. From the 2008 presidential election of Barack Obama in the US to the 2014 Indian general election, the use of digital media as a key tool of political communication has become the new trend (Castells, 2009). With time, not only has the usage of digital media as a tool of political communication improved and become stronger, but also, with the rise of misinformation, uncontrolled usage, and unequal web literacy, it has faced several critical problems (Morozov, 2011). An important aspect of using digital media for political communication is that the audience can freely communicate with the politicians and express their views about politics without fear of not being heard (Kumar, 2019). With its remarkably strong growth, this field also added some new words to its repository, all thanks to social media. Words like Fake News, Post Truth, Misinformation, Black Hat Propaganda, Spin, Information Warfare etc., which are either recently coined or came forth after taking a backseat for long are rampantly used nowadays (Lawson & Niblock, 2016). Controversies such as the election of Donald Trump, Cambridge Analytica and tension between western world and the Russian federation has made everyone rethink the power of social media in this arena (Faris & Culotta, 2017).

In the Indian context, the evolution of political communication can be divided into three phases. In the first phase, communication was used to sell relatively strong and stable political beliefs; as no strong opponent was there, people were largely inclined towards one thought, whereas in the second phase, a much more mobile/ undecided electorate was faced by the political parties, so there was a shift from two way/ face to face communication to mass communication mediums(Patel,2018). In the third phase, the emergence of choice and the availability of a plethora of information have created a confused and misinformed electorate in the country, so every political party is making the best use of technology to reach a wider audience for communicating not only on a party level but individually too, gathering feedback, and redesigning their strategies (Siapera, 2012). The exponential rise of digital social media and its usage in contemporary political communication in India has put forward new opportunities and challenges to what many have discarded as an intellectual impasse (Etling et al, 2017). Since the 2014 Lok Sabha elections in India, digital platforms have been a key part of how politicians talk to the public. In this election, the real candidate for Prime Minister from the Bhartiya Janata Party used the digital medium to interact with the people and create a discourse about his politics through this medium. From election campaign jingles to online townhouses, mobile apps for direct communication, and in many other unique ways, he used digital media, (O'Cass & Raltson,2016). After each election in India, digital media has become the most popular way for politicians to get their message out. All the other political parties, like the Aam Aadmi Party, Congress, Trinamool Congress, Samajwadi Party, and others, have been aggressively using digital media as a key tool of communication.

The dynamics of political communication are changing not only at the national level but state level. Interference from Central leadership, mudslinging, Twitter wars, and many more new age tactics have changed the game altogether (Kaun & Svensson,2017). Recently, India has undergone elections in four states, i.e., Gujrat, Himachal Pradesh, Tripura, and Meghalaya. Gujarat is one of the most developed states in India, and when Narendra Modi was the Chief Minister of the state, there was a huge push towards digital in every area. On the other hand, Himachal Pradesh includes a lot of villages in remote areas of the Himalaya where the digital reach is less. In Tripura, there is a significant section of the tribal population that lives in remote areas and might not have digital literacy. On the other hand, in Meghalaya, the population is highly divided between different tribes, classes, and religions, and maybe the usage of digital media is different based on many factors. Because of this, it's getting more and more important to study these states and talk about politics through digital media (Lilleker & Jackson, 2011). The diversity in development, nature of people, politics, culture, and other issues could have a direct impact on the usage of digital media (Rieder,2013). In these elections, a perfect blend of nativeness, traditional media, and digital media was witnessed, giving shape to the new age of political communication in India. In this research, we will look at how digital media was used as a key way to communicate in the recent state elections in India. The chosen four states, i.e. Gujarat, Himachal Pradesh, Tripura, and Meghalaya, are unique in their own way in terms of culture, demography, and politics. The study focused on these four state elections will not only bring out how digital media helps political parties across the spectrum develop political agendas and discourse but also explore the limitations of too much dependency on digital media.

The research will explore how digital media was used as a key tool of communication by all the political parties in these state elections, giving a new age to communication along with the new dimensions explored. The research will also explore how a successful digital media campaign can bring positive electoral results and whether there is a connection between a successful digital media campaign and electoral victory or not. This research would also look at the key components and parameters that political parties are considering for digital media campaigns while keeping the diverse electorate in mind.

OBJECTIVES

- To explore how digital media is used as a key tool of communication by all the political parties in these state elections.

· To investigate how effectively these political parties have used digital media in their political campaigns to establish the political discourse and set the agenda for public opinion.
· To explore how a successful digital media campaign can bring positive electoral results and whether there is a connection between a successful digital media campaign and electoral victory or not.
· To find the key components and parameters that political parties are considering for digital media campaigns while keeping the diverse electorate in mind.
· To look at whether political parties change how they use digital media in election campaigns based on the people they are trying to reach.
· To understand the different gaps in the use of digital media as a key tool of political communication and the scope of improvement for politicians with respect to the use of digital media.

RESEARCH METHODOLOGY

Taking the major four state elections in consideration, Content analysis of the tweets, post and Public Relations activities of the major parties will be undertaken. Content analysis of the tweets, post and Public relations activities of the major political leaders in these regions will also be studied. Once all the material is gathered, conclusions will be drawn thereafter.

FINDINGS

During the study, it was found that all the major political parties, like the Bharatiya Janata Party, Congress, Trinamool Congress, Aam Aadmi Party, and others, extensively used digital media for political communication. The study observed that in the given states, the political parties have a significant social media presence on platforms like Facebook, Twitter, and Instagram. It was also noted that all the political parties have specific social media pages for specific states that led the campaigns.

Table 1. Social media followers of three main political parties in Gujarat

Political Party	Facebook	Twitter	Instagram
BJP	36,00,000	15,00,000	8,95,000
Congress	701,000	1,77,900	68,200
AAP	576,000	1,51,700	1,15,000

Table 2. Social media followers of three main political parties in Tripura

Political Party	Facebook	Twitter	Instagram
BJP	3,82,000	1,03,500	8,279
Congress	78,000	62,300	NA
TMC	62,000	24,000	NA

Table 3. Social media followers of three main political parties in Himachal

Political Party	Facebook	Twitter	Instagram
BJP	3,023,000	2,11,200	24,600
Congress	3,34,000	61,300	25,400

According to the research, all of the major political parties in a given state have used social media for electoral campaigning. In the state of Gujarat, the electoral contest was primarily among three prominent political parties, namely the Bharatiya Janata Party (BJP), Indian National Congress (INC), and Aam Aadmi Party (AAP). The research findings indicate that the three political parties under investigation exhibit a noteworthy presence across the social media platforms of Facebook, Twitter, and Instagram. The Bharatiya Janata Party (BJP) emerged as the leading political party in terms of social media following across the three major platforms, namely Facebook, Twitter, and Instagram. The Facebook pages of political parties in Gujarat, namely BJP, Congress, and AAP, have varying numbers of followers. Specifically, BJP Gujarat's page has approximately 36,00,000 followers, while Congress and AAP have 701,000 and 576,000 followers, respectively. According to data collected from Twitter, the BJP Gujarat has amassed approximately 1.5 million followers, while the Congress and AAP have garnered 177,900 and 151,700 followers, respectively. The BJP Gujarat has a significantly higher number of followers on Instagram, with approximately 8,95,000 individuals following their account. In comparison, the Congress has 68,000 followers and the AAP has 1,15,000 followers on the same platform (Data collected from Pew Research Center, 2018).

The research study reveals that in comparison to larger states, the utilization of social media for campaign purposes was comparatively lower in states such as Tripura and Meghalaya. In this scenario, political parties have predominantly utilized Facebook and Twitter as a means of political communication. It has been observed that in the state of Tripura, the political parties Congress and Trinamool Congress do not possess official Instagram pages. The findings of the study indicate that the Bharatiya Janata Party (BJP) and the Indian National Congress (INC) have demonstrated active usage of all three major social media platforms in the state of Himachal Pradesh, despite its relatively smaller geographic and demographic size.

Personalized Communication and Use of Vernacular Language

The research indicates that social media platforms have been utilized by all political parties during election campaigns to facilitate personalised communication with potential voters. It can be observed that various political parties such as the BJP, Congress, Aam Aadmi Party, Trinamool Congress, and others have employed the use of vernacular language on their respective state-specific social media platforms. The linguistic medium of communication utilized by political parties in Tripura on their social media platforms is primarily Bengali. In terms of the primary pages or profiles of political parties, they tend to exhibit a preference for either English or Hindi. Likewise, in the state of Gujarat, all political factions have employed the Gujarati language in their social media communications and promotional materials.

According to the study's findings, the political parties involved in the social media campaign primarily emphasized state-centric issues in their use of vernacular language. Political parties tend to emphasise issues that are centred around the state in their social media campaigns (Figures 1, 2, and 3). The communication strategies employed by policymakers involve the identification and prioritization of local issues, the promotion of regional accomplishments, and the emphasis on policies that are applicable to the particular state or region. The utilisation of this approach enables political parties to customise their communication strategies in order to effectively connect with the local electorate and cater to their distinct requirements and aspirations.

Figure 1. Tripura BJP has highlighted several schemes of the Tripura government under them

Figure 2. Congress party in the Gujarati language has shared graphics on Facebook with the state-specific promises for Gujarat

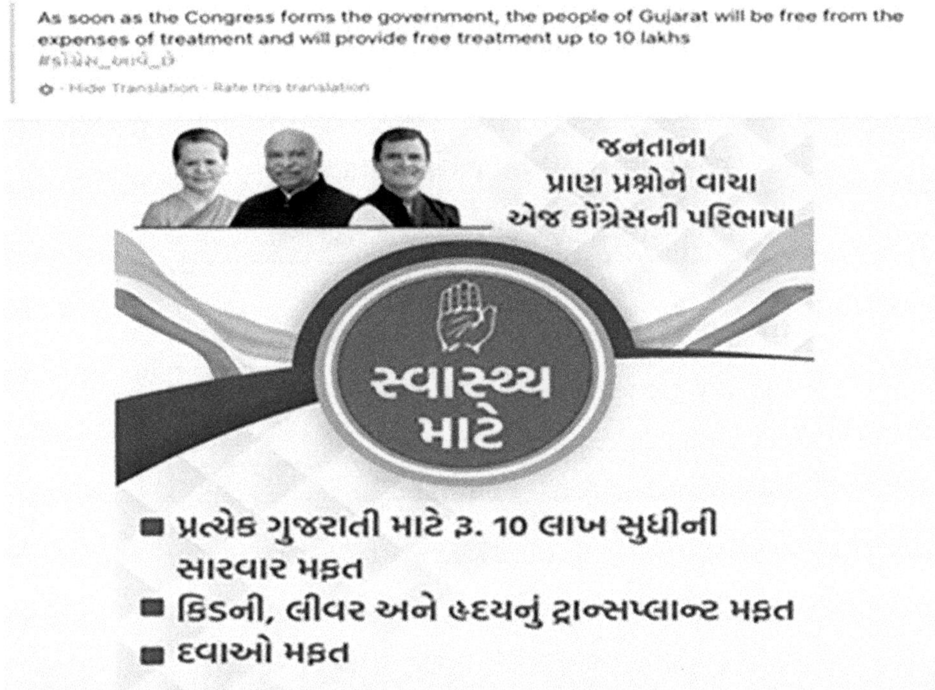

Figure 3. Graphics used by the AAP in Gujarat in Gujarati to apprise the voters about a decision taken by the state BJP

The research indicates that political parties attempted to integrate their campaigns into the cultural fabric of the local community by utilising vernacular languages. The utilisation of idioms, phrases, and references that are recognisable to the local population by political parties has been observed to enhance engagement and comprehension. The research findings also indicate that political parties strategically utilise vernacular language as a means to access and mobilise regional identity and pride. Political parties have utilised social media posts to improve their outreach by addressing fundamental socio-cultural and economic issues. The utilisation of vernacular languages in social media campaigns facilitated the political parties' ability to establish a strong connection with the electorate, tackle regional concerns, and exploit cultural significance.

Figure 4. TMC has shared a picture of one of its office bearers' son's picture claiming that he was allegedly beaten up by the BJP. Here the TMC claims that citizens of Tripura are not safe under the BJP

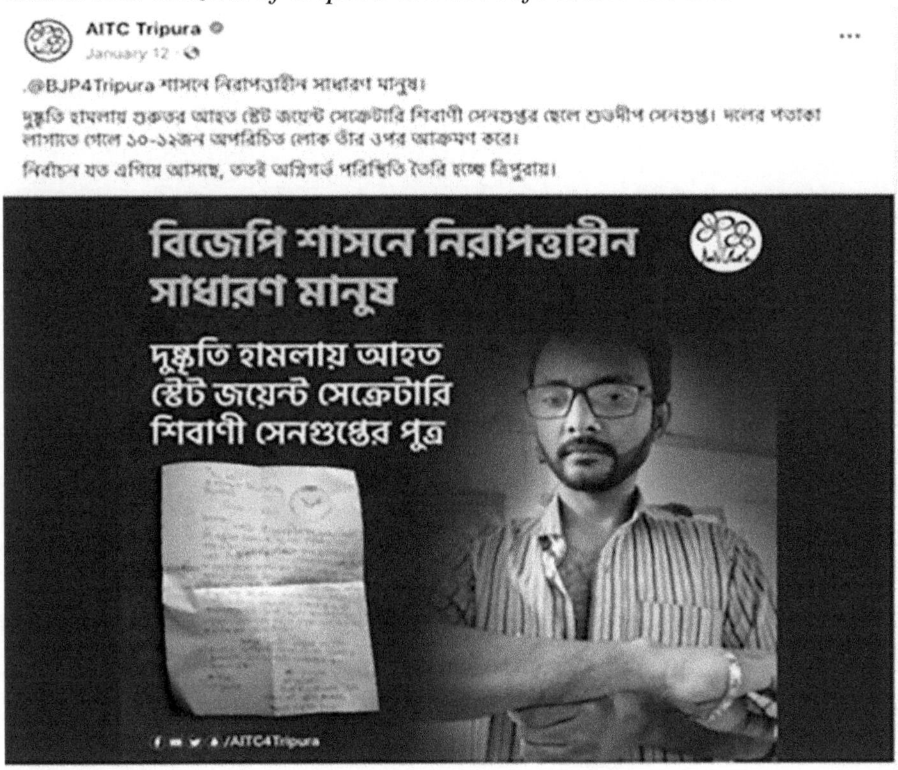

Below The Belt Attacks

The research indicates that political parties have resorted to personal attacks and derogatory comments while targeting their opponents on social media platforms, particularly by employing vernacular language. In instances of such attacks, political parties tend to divert their attention away from constructive issues at hand and instead concentrate on individuals.

Figure 5. Himachal Pradesh campaign the BJP focused targeted the Nehru-Gandhi family to project the success of schemes under Prime Minister Narendra Modi

In Figure 5, the BJP opted to portray the achievements of the Narendra Modi administration indirectly by drawing a comparison to the Nehru-Gandhi dynasty. The image under consideration depicts success schemes that are not directly linked to the Gandhi family. However, during the election campaign on social media, the BJP utilized personalized attacks against the Congress through the deployment of such graphics.

Figure 6. The Congress claimed that infighting within the BJP during Himachal Pradesh campaign

In a comparable vein, the use of a cartoon featuring Prime Minister Narendra Modi by the Congress in Figure 6 serves to underscore the presence of internal discord within the BJP in Himachal Pradesh. The selection of an image for the Prime Minister's cartoon on social media represents a form of nuanced personal criticism directed towards opposition political parties.

Use of Campaign Songs and Different Types of Videos

The findings of the study indicate that the use of campaign songs and videos has garnered noteworthy traction in state-level elections in India. The use of audiovisual mediums by political parties has been acknowledged across the political spectrum as an effective means of engaging voters, establishing a brand image, and communicating their message. As a result, many political parties have incorporated such songs and videos into their campaign strategies for the current elections (Figures 7 and 8).

The study observed that campaign songs and videos, encompassing video messages, live streaming of rallies, speeches, and press conferences, and individual videos that concentrate on state elections, possess a potent emotional appeal. Political parties have perceived this appeal as advantageous in fostering rapport with voters. Various mediums employ a combination of music, visuals, and storytelling techniques to elicit emotional responses, including but not limited to patriotism, hope, unity, and empathy. Political parties endeavour to create a lasting impact on the electorate by evoking emotional responses.

The research conducted indicates that campaign songs and videos serve as a medium for political parties to construct a persuasive storyline or narrative. They may emphasise the party's successes, the development of the leader, or the ambitions of the populace. The purpose of these narratives is to exert an impact on the collective consciousness of the public, mould their viewpoints, and cultivate a favourable impression of the political party and its leadership. The study observed that campaign songs and videos tend to align with the preferences and inclinations of the younger age group, which exhibits a greater propensity for online media consumption.

The research findings indicate that campaign songs and videos serve as a medium for effectively communicating important messages, slogans, or policy priorities in a succinct and influential way. By utilising memorable lyrics, visuals, and repetition, political parties can effectively reinforce their core messages and enhance the likelihood of voter recall.

Figure 7. BJP's campaign song and video in Tripura

Figure 8. AAP's campaign music video for Gujarat

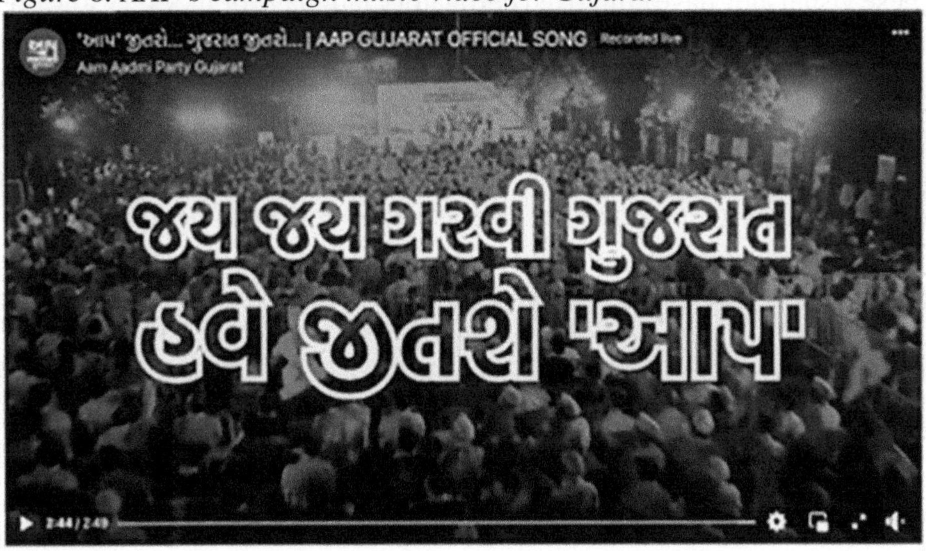

Use of Graphics and Cartoons

The research revealed that political parties have adopted a growing trend of utilising graphics and cartoons on social media platforms as a component of their electoral strategies across all four state elections. The visual components were instrumental in attracting notice, transmitting information, and interacting with the electorate. The study findings indicate that the utilisation of graphics and cartoons by political parties facilitated the visual storytelling process, allowing for the representation of intricate concepts and messages in a simplified and relatable manner. Political parties employed various visual aids, characters, and narratives to effectively convey their policies, accomplishments, and pledges.

The research revealed that political parties frequently utilised comical graphics and cartoons, such as memes and satirical material, as a means of connecting with younger demographics. Memes are a form of communication that utilise clever captions and visuals to express political ideas or opinions in a humorous and playful way. The implementation of this approach has the potential to enhance the shareability and accessibility of political content. However, it has been noted that in numerous instances, these visual aids have employed derogatory language or mockery towards opposing viewpoints.

The research revealed that infographics were utilised by political parties across all four state elections as a means of visually presenting information, data, and statistics in a manner that was both aesthetically pleasing and comprehensible. Infographics serve as a tool to streamline intricate subjects, policy endeavours, or political accomplishments, thereby enhancing their accessibility to the general populace.

Figure 9. A graphic shared by the BJP Tripura unit on Facebook which is a cartoon of fight between CPIM and Congress

Figure 10. A graphic shared by the BJP in during the Gujarat election campaign

Figure 11. A graphic shared by the TMC during Tripura elections

FOCUS ON CENTRAL LEADERS AND NARRATIVES RATHER THAN LOCAL

The research findings indicate that the dominant political parties with a nationwide presence, such as the Bharatiya Janata Party and the Congress, have primarily prioritised central leaders and a centralised narrative over the promotion and emphasis of local leadership and issues. As per the findings of an article published on thecnbctv18.com, ""“Only 15 percent of the tweets posted by Congress' official Twitter handle during November

21-27 were on the Gujarat polls. Out of 280 tweets posted during these seven days, only 42 were related to the party's campaigns and party leaders' rallies in Gujarat."

The data also reveals that the Bharatiya Janata Party (BJP) has strategically employed the triumph of the central government under the leadership of Narendra Modi as the primary electoral discourse in all the state-level elections. The present study reveals that the dominant political factions have intricately interwoven concerns pertaining to state-level affairs and the management of public affairs with those of national significance. The present discourse on Elections, despite the parties' attempts to construct a vernacular-based narrative on social media, has remained predominantly centralised in nature. The present study highlights the prominence of national leaders such as Prime Minister Narendra Modi, Home Minister Amit Shah, Congress leader Rahul Gandhi, and TMC supremo Mamata Banerjee in the social media campaigns undertaken by their respective political parties. The investigation has additionally ascertained that, as opposed to the social media accounts of the regional factions of these political organisations, the centralised storylines were predominantly constructed within the central party accounts on social media.

The study also discovered that when it comes to assembly elections, political parties, particularly national parties like the Bharatiya Janata Party and Congress, have a growing tendency to focus on national issues and national leaders rather than local leaders and issues. It was discovered that the Congress Party had paid little attention to the assembly elections in any of the states, including Gujarat, Himachal Pradesh, Meghalaya, and Tripura, focusing instead primarily on its senior leader, Rahul Gandhi, and the success of the Bharat Jodo Yatra. The image of a tweet (Figure 12) shared by the Congress party during the Gujarat assembly elections is shown below. It highlights the fact that the party chose to keep its focus on Rahul Gandhi, the national leader of the Congress, rather than discuss local problems and issues pertaining to Gujarat. In all the states examined here, this was the pattern for Congress.

Figure 12. Tweet shared by Congress during Gujarat election campaign

Figure 13. A picture of Home Minister Amit Shah and his message before Gujarat election

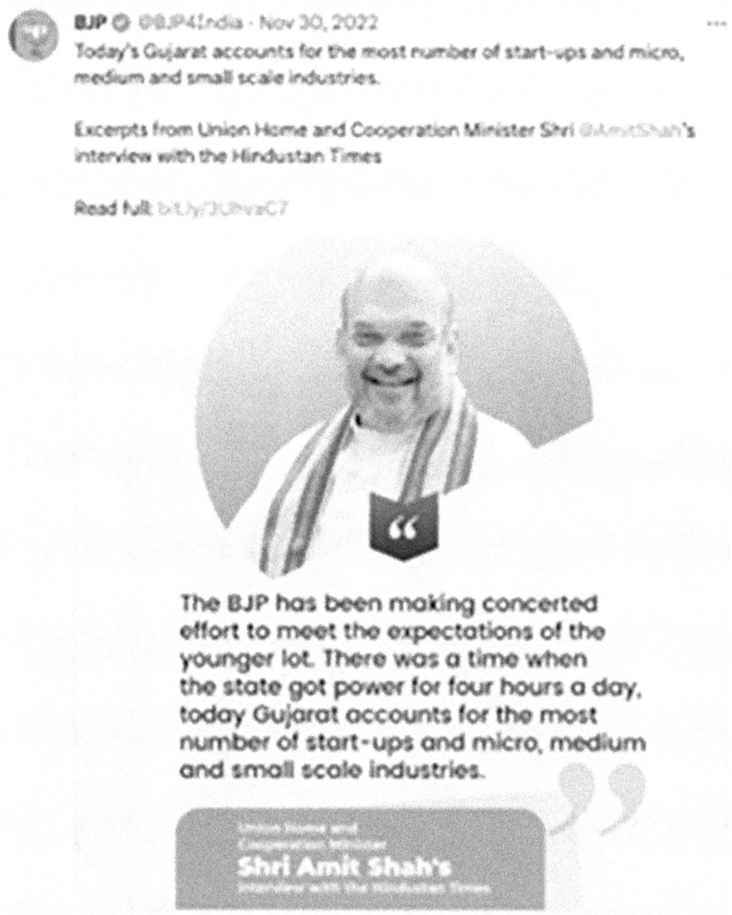

The study also revealed that the Bharatiya Janata Party preferred to discuss the core programmes of the BJP-controlled central government. It was discovered that the BJP leveraged the success of the national programmes in the state election campaigns in all of these elections. Even though the BJP was in power in states like Gujarat, Tripura, and Himachal Pradesh at the time of the in the elections, the party preferred to discuss the success of central government initiatives over those of the corresponding state governments.

CONCLUSION

The case study on the changing dynamics of political communication in the digital era in India reveals the significant impact of digital platforms on political campaigns, public discourse, and citizen engagement. The findings highlight the transformative nature of the digital era and its implications for political communication strategies and the overall democratic process in India.

Firstly, the study underscores the evolution of political communication in India, from traditional methods to the advent of digital platforms. The rise of social media platforms has provided political parties and leaders with new avenues to connect with voters, disseminate information, and shape public opinion. The speed and reach of digital communication have revolutionized political campaigns, enabling real-time engagement with constituents.

Secondly, the study highlights the strategies employed by political parties and leaders in leveraging digital platforms. Social media has become a key battleground for political messaging, with parties using data analytics, targeted advertising, and community-building techniques to mobilize support. The ability to reach specific demographics and engage with voters on a personalized level has reshaped the political landscape in India.

Thirdly, the study emphasizes the role of citizens in the digital era of political communication. The increased accessibility to political information has empowered citizens to participate actively in public discourse. Social media platforms serve as spaces for citizens to voice their opinions, engage in political discussions, and influence public opinion. Digital activism and crowdsourcing have emerged as powerful tools for citizens to mobilize for social and political causes. However, the case study also highlights several challenges and concerns associated with political communication in the digital era. The proliferation of misinformation and fake news poses a threat to the integrity of political discourse. Privacy and data security concerns are critical issues that need to be addressed to ensure the responsible use of digital platforms. Additionally, the amplification of polarization and the creation of echo chambers present challenges for fostering healthy democratic dialogue.

In conclusion, the case study on India demonstrates the transformative impact of the digital era on political communication. The findings underscore the need for political parties, leaders, and policymakers to adapt to the changing dynamics and harness the potential of digital platforms effectively. Responsible and ethical use of digital communication, coupled with robust regulations, is crucial to ensure the integrity and inclusivity of political discourse in the digital era. By understanding the changing dynamics of political communication in India, this case study provides valuable insights and recommendations for stakeholders to navigate the digital landscape and foster a vibrant and participatory democracy. The future of political communication in India lies in embracing the opportunities presented by the digital era while addressing the challenges it brings forth.

REFERENCES

Alam, M. S., & Das, S. K. (2019). The role of social media in political communication in India. In Alam, M. S., Saha, B., & Roy, A. K. (Eds.), *Social media and political communication in India* (pp. 1–14). Springer.

Bakir, V., & McStay, A. (2018). *Fake news and the political communication landscape*. Palgrave Macmillan.

Castells, M. (2009). *Communication power*. Oxford University Press.

Chadwick, A. (2017). *The hybrid media system: Politics and power*. Oxford University Press.

Etling, B., Faris, R., Howard, P. N., & Nielsen, R. K. (2017). *Computational propaganda: Political parties, politicians, and political manipulation on social media*. Oxford University Press.

Faris, R., & Culotta, A. (2017). The spread of news on Facebook. In *Proceedings of the 2017 ACM SIGMETRICS/IFIP Performance Evaluation Conference* (pp. 311-322). New York: ACM.

Kaun, M., & Svensson, J. (2017). *Digital media and political communication in India: Framing the contemporary public sphere*. Oxford University Press.

Kumar, R. (2019). *The impact of social media on political communication in India*. Sage Publications.

Lawson, S., & Niblock, S. (2016). *Political communication in the digital age*. Routledge.

Lilleker, D. G., & Jackson, N. A. (2011). *New media and political campaigning*. Sage Publications.

Morozov, E. (2011). *The net delusion: The dark side of Internet freedom*. Allen Lane.

Norris, P. (2000). *A new paradigm for digital democracy*. Cambridge University Press.

O'Cass, A., & Ralston, L. (2016). The political use of social media: A conceptual framework. In O'Cass, A., & Ralston, L. (Eds.), *The political use of social media* (pp. 1–18). Routledge.

Patel, P. (2018). *The digital elections: How social media is changing political communication in India*. Routledge.

Pew Research Center. (2018). *Social media use in India 2018*. Pew Research Center.

Rieder, B. (2013). *Digital politics: How technology is reshaping political participation*. Oxford University Press.

Siapera, K. (2012). *The political economy of search engines: An analysis of Google*. Palgrave Macmillan.

Chapter 3
Social TV and Nigerian Youth Culture Online:
Literature, Religiosity, and Preliminary Findings

Adamkolo Mohammed Ibrahim
https://orcid.org/0000-0003-1662-7054
University of Maiduguri, Nigeria

ABSTRACT

Social TV is the use of social media and digital technologies to enhance the TV viewing experience. However, there is a lack of research on social TV in Nigeria. This chapter aims to fill this gap by providing a critical analysis of social TV and audience interaction in Nigeria with some preliminary findings. A modified UTAUT2 was used as the theoretical framework, integrating technology awareness and religiosity as additional predicting and moderating variables respectively. A survey among Nigerian university students reveals that performance expectancy, social influence, hedonic motivation, price value, habit and technology awareness significantly predict use intentions, with extrinsic religiosity strengthening the relationship between behavioural intention and actual usage. The chapter also explores the potential and implications of social TV for content production, civic engagement, and cultural diversity in Nigeria. The chapter concludes by summarising the findings and contributions, and highlighting the limitations and directions for future research.

DOI: 10.4018/978-1-6684-9356-4.ch003

INTRODUCTION

Television (TV) is one of the most influential and pervasive mass media in the world. It has the power to inform, entertain, educate and persuade millions of viewers across different cultures and contexts. However, the traditional TV consumption paradigm, which is based on the linear and passive delivery and reception of TV content, is undergoing a radical transformation in the digital era. The emergence and proliferation of social media and digital technologies have enabled new forms of TV consumption that are interactive, participatory and social. This phenomenon is known as social TV, which can be defined as "the use of communication and social interaction features in relation to television content, or the use of television content to fuel social interaction" (Cesar et al., 2008, p. 1).

Social TV is not a new concept, as TV viewers have always engaged in social interactions around TV content such as talking to co-viewers, calling friends or joining fan clubs (Harboe et al., 2008). However, what is new and distinctive about social TV in the digital era is the use of social media platforms and devices such as Facebook, Twitter (X), smartphones and tablets, to enhance and extend the social TV experience. These platforms and devices allow TV viewers to access, share, comment, rate and create TV content anytime and anywhere, as well as to connect and interact with other TV viewers, TV producers and TV celebrities in real time or asynchronously (Doughty et al., 2012). Social TV thus represents a paradigm shift in TV consumption, as it transforms TV viewers from passive and isolated consumers to active and connected participants.

The phenomenon of social TV has attracted the attention of researchers from different disciplines such as media studies, communication studies, computer science and psychology. However, most of the existing studies on social TV have focused on the Western context, especially the United States and Europe, where social TV is more prevalent and advanced. There is a lack of research on social TV in the African context, especially in Nigeria, which is the largest economy and the most populous country in Africa, as well as one of the fastest-growing media markets in the world (DataReportal, 2022). Nigeria has a vibrant and diverse TV industry, with over 200 TV stations, both public and private that offer a variety of local and international TV content such as news, sports, entertainment and education (ThisDayLive, 2021). Nigeria also has a large and active social media user base, with over 31 million social media users, representing 14.3% of the total population, as of January 2022 (DataReportal, 2022). These factors suggest that social TV has a great potential and relevance in Nigeria, as TV viewers can use social media to enhance their TV consumption and interaction.

However, there is a lack of empirical and theoretical understanding of social TV and its implications in Nigeria. Therefore, the main objective of this chapter is to fill this gap by providing a comprehensive and critical analysis of social TV and audience interaction in Nigeria, and how it represents a paradigm shift in TV consumption (Ibrahim et al., 2024). The chapter is structured as follows: The first section reviews the historical and theoretical background of TV audience research in Nigeria, and discusses the influence of social media and digital technologies on the formation and behaviour of TV viewers in Nigeria. The second section identifies the different types of audience engagement and their measurement methods in Nigeria, and examines the various strategies and tools of social TV to promote audience participation and interaction in Nigeria. The third section explores the creative and innovative potential of social TV for content production and distribution in Nigeria, and illustrates the diverse and robust forms of communication and expression enabled by social TV in Nigeria. The fourth section assesses the implications of social TV for media literacy, civic engagement, and cultural diversity in Nigeria. The final section summarises the main findings and contributions of the chapter, highlights the limitations and directions for future research, and reflects on the significance and implications of social TV for the TV consumption paradigm in Nigeria.

The chapter adopts the Unified Theory of Acceptance and Use of Technology 2 (UTAUT2) as the theoretical framework to explain and predict the factors that influence the acceptance and use of social TV in Nigeria. The UTAUT2 model, developed by Venkatesh et al. (2012), is an extension of the original UTAUT model, which was proposed by Venkatesh et al. (2003) to integrate the key constructs from eight previous technology acceptance models. The UTAUT2 model proposes that seven main determinants affect the user's intention and use of technology: performance expectancy, effort expectancy, social influence, facilitating conditions, hedonic motivation, price value and habit. These determinants are moderated by four variables: gender, age, experience and behavioural intention. The UTAUT2 model has been widely applied and validated in various contexts of consumer technology adoption such as mobile internet, online shopping and e-learning. However, the UTAUT2 model may not fully capture the specific characteristics and influences of the Nigerian market, especially the role of religiosity in influencing the user's attitude and behaviour towards technology. Religiosity is defined as the degree of commitment, involvement or adherence to a certain religion or religious beliefs (Allport & Ross, 1967). Religiosity can affect the user's perception and evaluation of technology, as well as the social norms and expectations that guide the user's behaviour. For example, a study by Oyero and Salawu (2014) found that religiosity influenced the user's acceptance of mobile payment technology in Nigeria, as it affected the user's trust, perceived risk and perceived usefulness of the technology. Another study by Karamat Baig et al. (2021) found that religiosity moderated the effect of performance expectancy, effort expectancy, social influence and facilitating conditions on the user's intention to use mobile payment technology in traditional markets in Indonesia.

Therefore, to understand the adoption of social TV in Nigeria better, it is necessary to integrate technology awareness and religiosity into the UTAUT2 model as predicting and moderating variables respectively, in addition to gender, age, experience, andreligiosity with behavioural intention and use behaviour as the dependent variables. This modified UTAUT2 model can help to explain how technology awareness and religiosity affects the user's intention and use of social TV, as well as how it interacts with the other determinants and moderators of the model. This modified UTAUT2 model can also help to identify the challenges and opportunities of social TV adoption in Nigeria, and to provide recommendations for policy makers, practitioners and researchers who are interested in promoting and facilitating the adoption of social TV in Nigeria (Ibrahim et al., 2024).

The study is significant as it will contribute to the existing literature on social TV and audience interaction, as well as to the understanding of the TV consumption paradigm in Nigeria and beyond. The study will, first, provide a comprehensive and critical analysis of social TV and audience interaction in Nigeria, which is a largely under-researched and under-represented context in the global media landscape. Second, the study will explore the creative and innovative potential of social TV for content production and distribution in Nigeria, which is a dynamic and diverse media market with a rich and vibrant TV culture. Third, the study will assess the implications of social TV for media literacy, civic engagement, and cultural diversity in Nigeria, which are important and relevant issues for the TV industry and the society in Nigeria and other developing countries. Fourth, the study will identify the challenges and opportunities of social TV for the TV industry and the society in Nigeria, which can inform and inspire policy makers, practitioners, educators and researchers to develop and implement effective and sustainable social TV initiatives and interventions. Fifth, the study will reflect on the significance and implications of social TV for the TV consumption paradigm in Nigeria and beyond, which can provide new understanding and perspectives for the future of TV and media in the digital era.

It should be noted that this study provides a robust review of the literature with a summary of key preliminary findings of the empirical study. The summary of the findings are included in this chapter following editorial recommendations by through blind peer reviews. Hence, the addition of the 'Methods' section as well.

Aim and Objectives of the Study

The aim of the chapter is to provide a comprehensive and critical analysis of social TV and audience interaction in Nigeria, and how it represents a paradigm shift in TV consumption. The specific objectives of the study are categorised into two – objectives 1 – 7 focus on the review of the literature, while 6 – 8 focus on the preliminary findings of the empirical study). The objectives are to: 1) assess how social media and digital technologies influence the formation, behaviour and practices of TV audiences in Nigeria, with particular emphasis on the factors that influence their social TV preferences; 2) investigate the various modes of audience engagement and the methods employed to measure them in Nigeria, including the strategies and tools used to promote participation and interaction; 3) explore the creative potential of social TV for content production and distribution; 4) evaluate implications of social TV for media literacy, civic engagement and cultural diversity; 5) identify the attendant challenges and opportunities for the TV industry and society in Nigeria; 6) measure the frequencies of university students' actual usage of social TV; 7) examine the relationships between key determining factors of UTAUT2 and the students' usage of social TV; and 8) evaluate the moderating effects of gender, age, experience and religiosity on the relationship paths between key determinants of UTAUT2 and students' usage of social TV.

BACKGROUND

Social TV and the Transformation of Television Audience in Nigeria

Television (TV) is one of the most influential and pervasive mass media in Nigeria, reaching millions of viewers across different regions, cultures and languages. TV has the power to inform, entertain, educate and persuade the Nigerian audience on various issues and topics such as politics, sports, religion, culture and entertainment. However, the traditional TV consumption paradigm, which is based on the linear and passive delivery and reception of TV content, is undergoing a radical transformation in the digital era. The emergence and proliferation of social media and digital technologies have enabled new forms of TV consumption that are interactive, participatory and social. This phenomenon is known as social TV, which can be defined as "the use of communication and social interaction features in relation to television content, or the use of television content to fuel social interaction" (Cesar et al., 2008, p. 1).

To understand the influence and implications of social TV in Nigeria, it is important to review the historical and theoretical background of TV audience research in Nigeria and to examine how social media and digital technologies have influenced the formation and behaviour of TV viewers in Nigeria. It is also essential to analyse the challenges and opportunities of social TV for the TV industry and the society in Nigeria, and to explore the potential and prospects of social TV for the future of TV consumption in Nigeria.

Historical and Theoretical Background of TV Audience Research in Nigeria

Television TV audience research in Nigeria has a long and rich history, dating back to the inception of TV broadcasting in Nigeria in 1959. The first TV station in Nigeria, and indeed in Africa, was the Western Nigeria Television (WNTV), established by the Western Region government under the leadership of Chief Obafemi Awolowo. The WNTV was followed by the establishment of other regional TV stations such as the Eastern Nigeria Television (ENTV) in 1960, the Northern Nigeria Television (NNTV) in 1962 and the Midwest Television (MWTV) in 1964. These TV stations were mainly used as instruments of political propaganda and mobilisation by the regional governments, and had little regard for the needs and preferences of the TV audience. The TV audience research at this period was mainly based on the census data and the feedback from the viewers through letters and phone calls (Ojebode, 2010).

The situation changed in 1976, when the military government of General Olusegun Obasanjo established the Nigerian Television Authority (NTA) as the sole national TV network, and took over all the existing regional TV stations. The NTA was mandated to provide TV services that would reflect the national unity, diversity, and development of Nigeria and to cater for the interests and aspirations of the Nigerian TV audience. The NTA also initiated a more systematic and scientific approach to TV audience research, using various methods such as surveys, ratings, focus groups and content analysis, to measure the size, composition and preferences of the TV audience, and to evaluate the quality, influence and effectiveness of the TV programmes (Ojebode, 2010).

The NTA monopoly was broken in 1992, when the civilian government of President Ibrahim Babangida deregulated the broadcast sector, and allowed the entry of private TV stations such as the African Independent Television (AIT), the Channels TV and the Silverbird TV. These TV stations introduced more competition, diversity and innovation to the TV industry, and offered more choices and options to the TV audience. The TV audience research at this period became more sophisticated and competitive, as the TV stations employed various techniques such as ratings, surveys, focus groups and online feedback, to monitor and understand the

TV audience behaviour, and to attract and retain the TV audience loyalty and satisfaction (Ojebode, 2010).

The advent of the digital era in the late 1990s and early 2000s brought new challenges and opportunities to the TV industry and the TV audience research in Nigeria. The emergence and proliferation of social media and digital technologies such as the Internet, mobile phones, computers and tablets, enabled new forms of TV consumption that are interactive, participatory and social. These forms of TV consumption are collectively known as social TV, which can be seen as a convergence of TV and social media, or as an extension of TV through social media. Social TV allows the TV audience to access, share, comment, rate and create TV content anytime and anywhere, as well as to connect and interact with other TV viewers, TV producers and TV celebrities in real time or asynchronously. Social TV also provides new sources and methods of TV audience research such as social media analytics, online surveys, web tracking and user-generated content, to measure and understand the TV audience engagement, participation and interaction (Doughty et al., 2012).

Theoretical Background

The theoretical frameworks that can be used to study social TV and TV audience research in Nigeria are diverse and multidisciplinary, depending on the research questions and perspectives. Some of the possible theories are as follows. First, communication and media effects theories: These theories examine how TV messages influence the cognitive, affective and behavioural outcomes of TV viewers. For example, the selectivity, transactionality and conditionality paradigms (Severin & Tankard, 2001), the differential susceptibility to media effects model (Valkenburg & Peter, 2013) and the self-effects theory (Perse, 1996). Second, computer-mediated communication (CMC) theories: These theories explore how the characteristics of social media and digital technologies affect the communication processes and outcomes of TV viewers. For example, the social presence theory (Short et al., 1976), the media richness theory (Daft & Lengel, 1986) and the social information processing theory (Walther, 1992). Third, affordance theories: These theories focus on the potential actions and opportunities that social media and digital technologies offer to TV viewers. For example, the transactional affordance theory of social media uses (Sundar, 2008), the uses and gratifications theory (Katz et al., 1974) and the social cognitive theory (Bandura, 2001). Fourth, social identity and social influence theories: These theories investigate how TV viewers relate to and interact with others through social media and digital technologies. For example, the social identity theory (Tajfel & Turner, 1979), the social identity model of de-individuation effects (SIDE) (Lea & Spears, 1991) and the social influence theory (Cialdini & Goldstein, 2004). Fifth, technology Adoption Theories: These theories investigate how users accept and use new information technology and related technologes. For example, the technology adoption model (TAM), theory of reasoned action (TRA), theory of planned behaviour (TPB), diffusion of innovation theory (DIT), unified theory of acceptance and use of technology (UTAUT), etc. This study adopts the extended version of UTAUT (UTAUT2) and examined its application in the acceptance and use of social TV phenomenon.

The UTAUT2 Model as the Theoretical Understanding for this Chapter

To understand the adoption of new technology in Nigeria, especially in the context of mobile payment and fintech, it is useful to apply a theoretical framework that can capture the factors and processes that influence the user's intention and behaviour. One such framework is the Unified Theory of Acceptance and Use of Technology (UTAUT), which was developed by Venkatesh et al. (2003) to integrate the key constructs from eight previous technology acceptance models. The UTAUT model proposes that four main determinants affect the user's intention and use of technology: performance expectancy, effort expectancy, social influence and facilitating conditions. These determinants are moderated by four variables: gender, age, experience and voluntariness of use.

However, the UTAUT model was originally designed to explain the adoption of technology in organisational settings, not in consumer markets. Therefore, Venkatesh et al. (2012) extended the UTAUT model by adding three new constructs that are relevant for consumer behaviour: hedonic motivation, price value and habit. They also replaced voluntariness of use with behavioural intention as a moderator. This extended model is called

UTAUT2, and it has been applied to various contexts of consumer technology adoption such as mobile internet, online shopping and e-learning.

Modifying UTAUT2 Model by Integrating Religiosity as a Moderator Variable

However, UTAUT2 may not fully capture the specific characteristics and influences of the Nigerian market, especially the role of religiosity in influencing the user's attitude and behaviour towards technology. Religiosity is defined as the degree of commitment, involvement, or adherence to a certain religion or religious beliefs (Allport & Ross, 1967). Religiosity can affect the user's perception and evaluation of technology, as well as the social norms and expectations that guide the user's behaviour (Ibrahim et al., 2024). For example, a study by Oyero and Salawu (2014) found that religiosity influenced the user's acceptance of mobile payment technology in Nigeria, as it affected the user's trust, perceived risk and perceived usefulness of the technology. Another study by Karamat Baig et al. (2021) found that religiosity moderated the effect of performance expectancy, effort expectancy, social influence and facilitating conditions on the user's intention to use mobile payment technology in traditional markets in Indonesia.

This literature review examines awareness (in this context, technology awareness) as a multifaceted construct that encompasses knowledge, perception and understanding of new technological innovations. It argues that such awareness forms the foundation for technology adoption, as it not only informs the existence of individuals technology and potential benefits but also influences their interpretative lens through cultural, social and personal experiences, ultimately promoting a deeper comprehension of practical applications (Adekunle, 2020; Nielsen, 2021). Key insights from Rogers (2003), Davis (1989) and Venkatesh et al. (2003) underscore the importance of perceived usefulness, ease of use and social influences in driving acceptance.

Furthermore, the review highlights the critical relationship between technology awareness and subsequent adoption, particularly within the Nigerian context. Increased levels of awareness have been shown to correlate with higher ICT adoption (Rogers, 2003). In the rapidly emerging arena of social television – which merges traditional broadcasting with interactive social media – the significance of technology awareness is further emphasised. Evidence from studies in Nigeria, South Africa and the United States (e.g., Adekunle, 2020; Nielsen, 2021; SABC, 2020) demonstrates that a well-informed audience tends to exhibit greater engagement and satisfaction with social television platforms. Ultimately, the review advocates for the promotion of technology awareness as a vital means to stimulate wider adoption and enrich user experience across diverse media platforms.

Therefore, to understand the adoption of social TV in Nigeria better, it is necessary to integrate technology awareness and religiosity into the UTAUT2 model as the eight determining factor and fourth moderating variable respectively, in addition to the exiting seven key determinants and three moderating variables in the model. This modified UTAUT2 model can help to explain how technology awareness and religiosity influence the user's intention and use of social TV, as well as how it interacts with the moderators of the model.

ISSUES AND CONTROVERSIES IN SOCIAL TV ADOPTION

Influence of Social Media and Digital Technologies on User Behaviour in Nigeria

Social media and digital technologies have a significant influence on the formation and behaviour of TV viewers in Nigeria, as they enable new forms of TV consumption that are interactive, participatory and social. These forms of TV consumption are collectively known as social TV, which can be seen as a convergence of TV and social media, or as an extension of TV through social media. Social TV allows the TV audience to access, share, comment, rate and create TV content anytime and anywhere, as well as to connect and interact with other TV viewers, TV producers and TV celebrities in real time or asynchronously. Social TV also provides new sources and methods of TV audience research such as social media analytics, online surveys, web tracking and user-generated content, to measure and understand the TV audience engagement, participation and interaction (Ibrahim et al., 2022, 2024; Winter et al., 2020).

The influence of social media and digital technologies on the formation and behaviour of TV viewers in Nigeria can be examined from different perspectives such as the individual, the social and the cultural perspectives. From the individual perspective, social media and digital technologies enable the TV viewers to have more control, choice and customisation over their TV consumption, as they can access, share, comment, rate and create TV content according to their preferences, interests and needs (Winter et al., 2020). Social media and digital technologies also enhance the TV viewers' enjoyment, satisfaction and loyalty, as they can engage in social interactions and feedback with other TV viewers, TV producers, and TV celebrities and express their opinions, emotions and identities. Social media and digital technologies also influence the TV viewers' cognition, affect and behaviour, as they can expose them to different types and sources of TV content, and influence their attitudes, beliefs and values. For example, a study by Oyero and Salawu (2014) found that social media platforms such as Facebook and Twitter, influenced the TV viewers' perception and evaluation of the 2012 fuel subsidy removal protest in Nigeria, as they provided alternative and diverse information and opinions on the issue.

From the social perspective, social media and digital technologies enable the TV viewers to form and maintain social relationships and networks with other TV viewers, TV producers and TV celebrities, and to participate in social activities and communities around TV content. Social media and digital technologies also facilitate the TV viewers' social identity and social influence, as they can identify with and influence others through their TV consumption and interaction (Ibrahim et al., 2024). Social media and digital technologies also affect the TV viewers' social norms and expectations, as they can create and follow social rules and conventions for their TV consumption and interaction. For example, a study by Oyero (2018) found that social media platforms such as WhatsApp and Facebook, enabled the TV viewers to form and sustain social groups and networks around TV programmes such as the Big Brother Naija reality show, and to engage in social interactions and activities such as voting, commenting and sharing, with other TV viewers.

From the cultural perspective, social media and digital technologies enable the TV viewers to access and create diverse and robust forms of TV content that reflect and represent their cultural identities, values and expressions. Social media and digital technologies also enhance the TV viewers' cultural awareness and diversity, as they can expose them to different cultures and perspectives through TV content, and promote intercultural dialogue and understanding with other TV viewers. Social media and digital technologies also affect the TV viewers' cultural norms and practices, as they can challenge and change the existing cultural patterns and meanings of TV consumption and interaction. For example, a study by Oyero and Oyesomi (2017) found that social media platforms such as YouTube and Instagram, enabled the TV viewers to access and create diverse and robust forms of TV content that reflected and represented their cultural identities, values and expressions such as the Nigerian comedy skits, music videos and fashion shows.

Social TV and the Enhancement of Audience Engagement

Audience engagement is a key concept in media studies, as it reflects the degree and quality of the relationship between the media and the audience. Audience engagement can be defined as "the process of involving audiences in the creation, dissemination and consumption of media content, as well as the emotional, cognitive and behavioural outcomes of such involvement" (Napoli, 2011, p. 5). Audience engagement can be measured by various indicators such as attention, participation, interaction, satisfaction and loyalty (Napoli, 2011; Winter et al., 2020).

Furthermore, audience engagement is a multidimensional concept that refers to the degree to which the audience is involved, interested and satisfied with a media product or service (Brodie et al., 2011). Audience engagement can be influenced by various factors such as the content, the context, the platform, the user characteristics and the social interactions (Napoli, 2014). Audience engagement can also have various outcomes such as increased attention, retention, loyalty, satisfaction, word-of-mouth and participation (van Doorn et al., 2010).

Social TV, as a phenomenon that combines TV and social media, has the potential to enhance audience engagement in various ways. Social TV can promote audience participation and interaction, as it allows the audience to access, share, comment, rate and create TV content anytime and anywhere, as well as to connect and interact with other TV viewers, TV producers and TV celebrities in real time or asynchronously. Social TV can also increase audience enjoyment, satisfaction and loyalty, as it enhances the audience's sense of community, social presence and emotional support, as well as their perception of the TV content's value, quality and relevance (Ibrahim et al., 2022, 2024; Winter et al., 2020).

In the next section, we will examine the various strategies and tools of social TV that can promote audience participation and interaction, and evaluate the effects of social TV on audience enjoyment, satisfaction and loyalty. We will use the UTAUT2 model as a theoretical framework to understand the factors and processes that influence the audience's intention and behaviour of using social TV.

Strategies and Tools of Social TV to Promote Audience Participation and Interaction

Social TV offers various strategies and tools to promote audience participation and interaction, such as:

- *Synchronous viewing*: This is the practice of watching television programmes at the same time as other viewers, either in the same physical location or in different locations connected by social media. Synchronous viewing can create a sense of co-presence, shared experience and social bonding among viewers, as well as increase the attention and excitement of watching (Harboe et al., 2008).
- *Live chats and comments*: These features allow the audience to communicate with each other and with the TV content creators and celebrities during or after the TV viewing. Live chats and comments can enhance the audience's sense of social presence, interactivity and feedback, as well as their enjoyment and satisfaction of the TV content. For example, a study by Winter et al. (2020) found that live chats and comments increased the audience's enjoyment and

satisfaction of watching a German crime show, especially when the comments were positive and relevant[2].

- *Second screen applications*: These are the applications that run on mobile devices or computers that provide additional information, content, or functionality related to the television programme. Second screen applications can enhance the viewing experience, enrich the content and increase the interactivity and personalisation of the programme (Cesar & Chorianopoulos, 2009).
- *User-generated content*: This is the content that is created and shared by the viewers themselves such as videos, images, memes, fan art, fan fiction or reviews. User-generated content can promote creativity, expression and empowerment among viewers, as well as generate buzz, word-of-mouth and viral effects for the media product or service (Jenkins, 2006).
- *Polls and quizzes*: These features allow the audience to express their opinions, preferences and knowledge about the TV content, as well as to see the results of other viewers. Polls and quizzes can enhance the audience's sense of participation, involvement and curiosity, as well as their enjoyment and satisfaction of the TV content. For example, a study by Khan and Khan (2020) found that polls and quizzes increased the audience's engagement and enjoyment of watching various TV programmes such as sports, entertainment and education. This study examined how the use of interactive media such as polls and quizzes enhanced the audience engagement and enjoyment of watching the popular Indian quiz show 'Kaun Banega Crorepati' (Who Wants to Be a Millionaire?). The study used a mixed-methods approach, combining survey data, focus group discussions and content analysis to explore the audience's perceptions, preferences and participation patterns. The study found that interactive media increased the audience's involvement, satisfaction and loyalty to the show, as well as their learning outcomes and social interactions.
- *Social sharing and user-generated content*: These features allow the audience to share their TV viewing experiences and create their own TV content on social media platforms such as Facebook, Twitter, YouTube and Instagram. Social sharing and user-generated content

can enhance the audience's sense of expression, creativity and community, as well as their enjoyment and satisfaction of the TV content. For example, a study by Oyero and Oyesomi (2017) found that social sharing and user-generated content increased the audience's engagement and enjoyment of watching diverse and robust forms of TV content that reflected and represented their cultural identities, values and expressions such as the Nigerian comedy skits, music videos and fashion shows.

Effects of Social TV on Audience Enjoyment, Satisfaction and Loyalty

Social TV can have positive effects on audience enjoyment, satisfaction and loyalty, as it can increase the audience's emotional, cognitive and behavioural outcomes of TV viewing. Audience enjoyment, satisfaction and loyalty are important indicators of audience engagement and media success. Some of these effects are:

- *Enjoyment*: This is the degree to which the audience experiences pleasure, fun or entertainment from the media product or service. Social TV can increase enjoyment by providing social presence, emotional contagion, humour and variety to the viewing experience (Banjo et al., 2015).
- *Satisfaction*: This is the degree to which the audience evaluates the media product or service as meeting or exceeding their expectations, needs or desires. Social TV can increase satisfaction by providing information, education, personalisation and feedback to the viewers (Lee et al., 2010).
- *Loyalty*: This is the degree to which the audience exhibits repeated or continued consumption, preference or advocacy of the media product or service. Social TV can increase loyalty by creating social identity, social influence, social capital and social norms among the viewers (Mora et al., 2011).

However, when the UTAUT2 model is applied as a conceptual framework (see Figure 1), these outcomes can be influenced by various factors. The factors are as outlined below.

Figure 1. The conceptual framework of the research based on the modified UTAUT2 model (Author, adapted with modifications from Venkatesh et al., 2012)

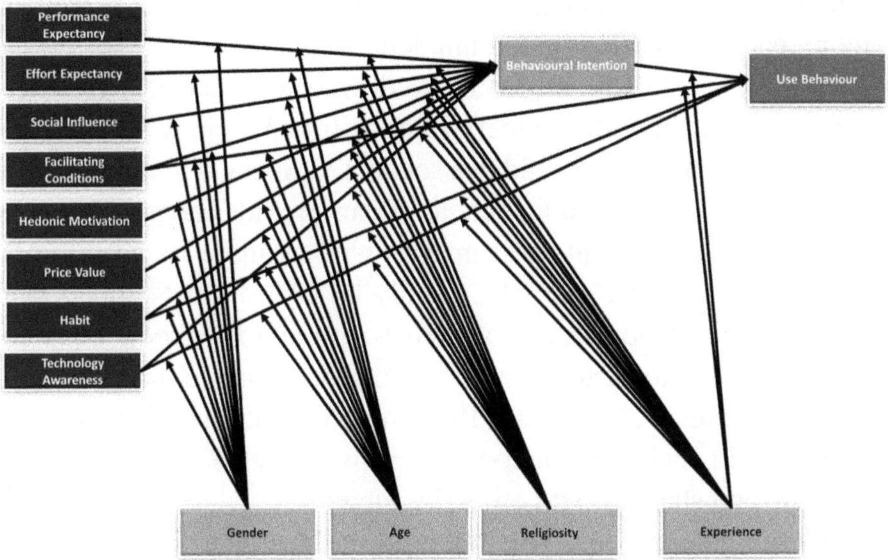

- *Performance expectancy*: This refers to the degree to which the audience believes that using social TV will help them achieve their goals such as obtaining information, entertainment or education. Social TV can increase the audience's performance expectancy, as it can provide them with more diverse, robust, and relevant TV content, as well as more opportunities for feedback and interaction. For example, a study by Oyero and Salawu (2014) found that social media platforms such as Facebook and Twitter, influenced the audience's perception and evaluation of the 2012 fuel subsidy removal protest in Nigeria, as they provided alternative and diverse information and opinions on the issue.

- *Effort expectancy*: This refers to the degree of ease associated with the use of social TV. Social TV can increase the audience's effort expectancy, as it can provide them with more convenient, accessible and user-friendly TV viewing experiences, as well as more guidance and support. For example, a study by Lee et al. (2010) found that social TV systems such as the Social TV Buddy, increased the audience's effort expectancy, as they provided them with more intuitive, personalised and adaptive TV viewing interfaces, as well as more recommendations and reminders.
- *Social influence*: This refers to the degree to which the audience perceives that important others such as friends, family or celebrities, believe that they should use social TV. Social TV can increase the audience's social influence, as it can provide them with more social cues, norms and pressures, as well as more social rewards and recognition. For example, a study by Oyero (2018) found that social media platforms such as WhatsApp and Facebook, enabled the audience to form and sustain social groups and networks around TV programmes such as the Big Brother Naija reality show, and to engage in social interactions and activities such as voting, commenting and sharing, with other TV viewers.
- *Facilitating conditions*: This refers to the degree to which the audience believes that the technical infrastructure and resources are available to support the use of social TV. Social TV can increase the audience's facilitating conditions, as it can provide them with more reliable, secure and compatible TV viewing systems and devices, as well as more technical assistance and troubleshooting. For example, a study by Karamat Baig et al. (2021) found that facilitating conditions such as the availability of internet access, mobile devices and mobile payment systems, influenced the audience's intention to use mobile payment technology in traditional markets in Indonesia.
- *Hedonic motivation*: This refers to the fun or pleasure derived from using social TV. Social TV can increase the audience's hedonic motivation, as it can provide them with more enjoyable, satisfying and rewarding TV viewing experiences, as well as more opportunities for self-expression, creativity and curiosity. For example, a study by Zillich (2020) found that co-viewing and social TV increased the

audience's hedonic motivation, as they provided them with more fun, excitement and stimulation, as well as more opportunities for social interaction, feedback and identity expression.

- *Price value*: This refers to the audience's cognitive trade-off between the perceived benefits and the monetary and non-monetary costs of using social TV. Social TV can increase the audience's price value, as it can provide them with useful, quality and relevant TV content, as well as more cost-effective, efficient and flexible TV viewing options. For example, a study by Banjo et al. (2015) found that social TV increased the audience's price value, as it provided them with more useful information, entertainment and socialisation, as well as more options to watch TV content for free or at a lower cost.
- *Habit*: This refers to the extent to which the use of social TV becomes an automatic behaviour that is triggered by situational cues. Social TV can increase the audience's habit, as it can provide them with more frequent, consistent and repeated TV viewing experiences, as well as more cues, triggers and reminders. For example, a study by Venkatesh et al. (2012) found that habit influenced the audience's use of mobile internet, as it became a routine and spontaneous behaviour that was triggered by contextual factors such as time, location and mood.

According to the UTAUT2 model, these factors can be moderated by four variables, namely gender, age and experience. However, we modified the model by incorporating religiosity as another moderator variable, the fourth, in this case:

- *Gender*: This refers to the biological and social differences between males and females. Gender can moderate the effects of social TV on audience engagement, as males and females may have different motivations, preferences and behaviours of using social TV. For example, a study by Venkatesh et al. (2012) found that gender moderated the effects of performance expectancy, effort expectancy, social influence and hedonic motivation on the audience's intention to use mobile internet, as males were more influenced by performance expectancy and effort expectancy, while females were more influenced by social influence and hedonic motivation.

- *Age*: This refers to the chronological age of the audience. Age can moderate the effects of social TV on audience engagement, as younger and older audiences may have different attitudes, perceptions and skills of using social TV. For example, a study by Winter et al. (2020) found that age moderated the effects of live chats and comments on the audience's enjoyment and satisfaction of watching a German crime show, as younger audiences were more influenced by live chats and comments, while older audiences were more influenced by the TV content itself.
- *Experience*: Experience refers to the amount of time that the audience has been using social TV. Experience can moderate the effects of social TV on audience engagement, as the audience who are more or less experienced may have different motivations, preferences and behaviours of using social TV. For example, a study by Venkatesh et al. (2012) found that experience moderated the effects of effort expectancy, social influence and hedonic motivation on the audience's intention to use mobile internet, as the audience who were more experienced were more influenced by effort expectancy and hedonic motivation, while the audience who were less experienced were more influenced by social influence.
- *Behavioural intention*: This refers to the degree to which the audience plans to use social TV in the future. Behavioural intention can moderate the effects of social TV on audience engagement, as the audience who intend to use social TV may have different motivations, expectations and behaviours of using social TV. For example, a study by Venkatesh et al. (2012) found that behavioural intention moderated the effects of facilitating conditions and habit on the audience's use of mobile internet, as the audience who intended to use mobile internet were more influenced by facilitating conditions, while the audience who did not intend to use mobile internet were more influenced by habit.
- *Religiosity*: This refers to the degree of commitment, involvement, or adherence to a certain religion or religious beliefs. Religiosity can moderate the effects of social TV on audience engagement, as the

audience who are more or less religious may have different perceptions, evaluations and behaviours of using social TV. For example, a study by Oyero and Salawu (2014) found that religiosity moderated the effects of performance expectancy, effort expectancy, social influence and facilitating conditions on the audience's acceptance of mobile payment technology in Nigeria, as the audience who were more religious were more influenced by performance expectancy, effort expectancy and social influence, while the audience who were less religious were more influenced by facilitating conditions.

Cultural and Social Influence of Social TV

Social TV, as a phenomenon that integrates television and social media, has a significant cultural and social influence, both locally and globally. In this section, we will explore how social TV affects the changes in cultural norms and values, as well as how social TV acts as a catalyst for social interaction and community building.

Exploration of Changes in Cultural Norms Due to Social TV

Cultural norms and values are the shared beliefs, expectations and behaviours that guide and regulate the actions and interactions of individuals and groups within a culture. Cultural norms and values are influenced by various factors such as history, geography, religion, language and media. Social TV, as a form of media, can affect cultural norms and values in both positive and negative ways. It can be used to spread knowledge, awareness and diversity, but it can also be used to spread stereotypes, misinformation and polarisation. A feeling of competition or comparison between different cultures can lead to a sense of cultural fragmentation or homogenisation (Ibrahim et al., 2022, 2024; Winter et al., 2020).

One of the ways that social TV can affect cultural norms and values is by reflecting and representing the existing or emerging cultural trends and issues in society. Social TV can provide a platform for showcasing and celebrating the diversity and richness of different cultures, as well as for raising and debating the challenges and opportunities that different cultures face. For example, in Nigeria, social TV can reflect and represent the cultural diversity and complexity of the country, which has over 250 ethnic groups and languages, and various religious and regional affiliations. Social TV can also reflect and represent the cultural trends and issues that Nigeria faces such as the impact of globalisation, urbanisation, migration and modernisation, on the traditional and contemporary aspects of Nigerian culture. Social TV can also reflect and represent the cultural aspirations and expressions of Nigerians such as their creativity, innovation and entrepreneurship, in various fields and domains such as music, fashion, art and literature (Oyero & Salawu, 2014).

Another way that social TV can affect cultural norms and values is by influencing the attitudes, opinions and behaviours of the audience and the society. Social TV can provide a source of information, education and entertainment, but it can also provide a source of persuasion, manipulation and socialisation. Social TV can influence the audience and the society by exposing them to different perspectives, values and lifestyles, as well as by reinforcing or challenging their existing beliefs, preferences and norms. For example, in Africa, social TV can influence the audience and the society by exposing them to different aspects of African culture, as well as to other cultures from around the world. Social TV can also influence the audience and the society by reinforcing or challenging their views and attitudes on various social issues such as gender equality, sexual health, mental health and social justice (Chuma, 2018).

CHALLENGES AND CONCERNS IN SOCIAL TV ADOPTION

Social TV, as a phenomenon that integrates television and social media, provides many benefits and opportunities for the audience, the producers and the society at large. However, it also poses many challenges and concerns, especially in terms of privacy, misinformation, algorithmic bias and challenges and opportunities of social TV for the TV industry and the society in Nigeria. In this section, we will explore these issues and their implications for social TV in Nigeria, Africa and elsewhere.

Privacy Considerations in Social TV

Privacy is a fundamental human right that protects the dignity, autonomy and freedom of individuals. Privacy is also a complex and contextual concept that depends on the expectations, preferences and values of different people, cultures and situations. Privacy is often threatened by the collection, processing and sharing of personal data, which can expose individuals to various risks such as identity theft, fraud, discrimination, harassment and surveillance. Privacy is particularly important and challenging in the context of social TV, as it involves the interaction of multiple actors, platforms and data sources, each with their own policies, practices and purposes.

One of the privacy challenges in social TV is the lack of awareness and control of the audience over their personal data and how it is used by the social media platforms and the television providers. The audience may not be fully informed or consented to the terms and conditions of the services they use, or the privacy settings and options they have. The audience may also not be aware of the types and amounts of data that are collected and shared by the platforms and the providers such as their viewing habits, preferences, opinions and interactions. The audience may also not be able to access, correct or delete their data, or to opt out of certain data practices such as targeted advertising, personalisation or profiling. The audience may also not be able to prevent or limit the access of third parties such as advertisers, marketers, researchers, or governments, to their data or to know how their data is used by these parties. These privacy challenges may undermine the trust and confidence of the audience in the social TV services, and may also expose them to various harms such as manipulation, exploitation or discrimination (Bucher & Helmond, 2018; Van den Broeck et al., 2015).

Another privacy challenge in social TV is the potential conflict and tension between the individual and the collective aspects of privacy. Social TV is a social and collaborative activity that involves the sharing and communication of information and opinions among the audience, the producers and the society. However, the audience may have different expectations and preferences regarding their privacy and the privacy of others. For example, some audience members may want to share their viewing experiences and opinions with others, while others may want to keep them private. Some audience members may want to see and interact with other viewers, while others may want to avoid or block them. Some audience members may respect and protect the privacy of others, while others may violate or expose it. These privacy challenges may create conflicts and tensions among the audience, and may also affect the quality and diversity of the content and communication in social TV (Bonsón et al., 2015; César et al., 2011).

To address these privacy challenges, social TV needs to adopt a user-centric and context-aware approach to privacy that respects and empowers the audience as the owners and controllers of their personal data, and that balances and harmonises the individual and the collective aspects of privacy. This approach may involve the following strategies:

- Providing clear and transparent information and consent mechanisms for the audience to understand and agree to the data practices of the social TV services, and to exercise their privacy rights and choices.
- Implementing privacy-by-design and privacy-by-default principles for the social TV services that minimise the collection and sharing of personal data, and that provide the highest level of privacy protection for the audience by default.
- Developing and applying privacy-enhancing technologies and techniques for the social TV services that protect the confidentiality, integrity and anonymity of the personal data, and that prevent or mitigate the risks of data breaches, leaks or abuses.
- Encouraging and facilitating privacy-aware and privacy-respectful behaviours and norms among the audience, the producers and the society that promote the awareness and appreciation of the value and importance of privacy, and that promote the trust and cooperation among the stakeholders (Bucher & Helmond, 2018; Van den Broeck et al., 2016).

Challenges and Opportunities of Social TV for the TV Industry and the Society in Nigeria

Social TV poses both challenges and opportunities for the TV industry and the society in Nigeria, as it transforms the TV consumption paradigm and the TV audience behaviour. The challenges and opportunities of social TV can be discussed from different perspectives such as the economic, the political and the ethical perspectives. From the economic perspective, social TV poses a challenge for the TV industry, as it creates more competition and fragmentation in the TV market, and reduces the TV viewers' dependence and loyalty to the traditional TV channels and programmes (Bucher & Helmond, 2018). Social TV also poses a challenge for the TV industry, as it requires more investment and innovation in the TV production and distribution, and more adaptation and integration of the social media and digital technologies. However, social TV also provides an opportunity for the TV industry, as it creates more demand and diversity for the TV content, and increases the TV viewers' engagement and participation. Social TV also provides an opportunity for the TV industry, as it provides new sources and methods of TV audience research, and new strategies and tools of TV marketing and advertising (Ibrahim et al., 2022, 2024).

From the political perspective, social TV poses a challenge for the society, as it creates more exposure and vulnerability to the TV content that may be biased, misleading or harmful, and may influence the TV viewers' opinions, attitudes and behaviours on various political issues and topics. Social TV also poses a challenge for the society, as it requires more regulation and governance of the TV content and the TV audience behaviour, and more protection and empowerment of the TV viewers' rights and responsibilities. However, social TV also provides an opportunity for the society, as it creates more access and diversity to the TV content that may be informative, educative or persuasive, and may enhance the TV viewers' awareness, knowledge and skills on various political issues and topics. Social TV also provides an opportunity for the society, as it facilitates more participation and interaction of the TV viewers with the political actors and institutions, and promotes more civic engagement and democracy (Bonsón et al., 2015; César et al., 2011).

From the ethical perspective, social TV poses a challenge for the society, as it creates more exposure and vulnerability to the TV content that may be unethical, immoral or illegal, and may violate the TV viewers' privacy, dignity or security. Social TV also poses a challenge for the society, as it requires more ethical awareness and education of the TV viewers, and more ethical standards and codes of conduct for the TV content and the TV audience behaviour. However, social TV also provides an opportunity for the society, as it creates more exposure and diversity to the TV content that may be ethical, moral or legal, and may respect the TV viewers' privacy, dignity or security. Social TV also provides an opportunity for the society, as it promotes more ethical reflection and dialogue of the TV viewers with the TV content and the TV audience behaviour, and promotes more ethical values and principles for the TV consumption and interaction (Oyero & Salawu, 2014).

CASE STUDIES AND EXEMPLARS OF SOCIAL TV MODELS

In this section, we will present some case studies of successful implementations of social TV strategies in Nigeria, Africa and elsewhere, as well as some lessons learned from notable failures or challenges in Nigeria, Africa and elsewhere. These case studies will illustrate the potential and the pitfalls of social TV for enhancing audience engagement, creating new forms of content and communication, and promoting media literacy, civic engagement and cultural diversity.

Successful Implementations of Social TV Strategies in Nigeria, Africa and Elsewhere

One of the successful implementations of social TV strategies in Nigeria is the Big Brother Naija reality show, which is a spin-off of the international Big Brother franchise. The show features a group of contestants who live together in an isolated house for several weeks, while being constantly monitored by cameras and microphones. The contestants compete for a cash prize and other rewards, while facing various tasks, challenges and evictions. The show is broadcasted live on television and online platforms, and engages the audience through social media platforms such as Twitter, Facebook, Instagram and YouTube. The audience can participate and interact with the show by voting for their favourite contestants, commenting on the events, sharing their opinions, creating and consuming user-generated content and joining online communities and groups. The show has been a huge success in Nigeria, attracting millions of viewers and followers, generating massive social media buzz and engagement and creating numerous celebrities and influencers. The show has also been praised for its cultural diversity and representation, as it showcases the different ethnic, linguistic, religious and regional backgrounds of the contestants and the audience. The show has also been credited for its educational and social impact, as it raises awareness and discussion on various social issues such as gender equality, sexual health, mental health and social justice (Oyero, 2018; Oyero & Oyesomi, 2017).

Another successful implementation of social TV strategies in Africa is the Jamii Forums, which is a popular online platform that hosts various discussions and debates on various topics such as politics, economy, culture, sports and entertainment. The platform also features live streaming and live blogging of various events such as elections, protests, sports matches and concerts. The platform engages the audience through social media platforms such as Twitter, Facebook, WhatsApp and YouTube. The audience can participate and interact with the platform by posting comments, questions, opinions and information, as well as by sharing, liking and rating the content. The platform has been a success in Africa, attracting millions of users and followers, generating massive social media traffic and engagement and creating numerous influencers and activists. The platform has also been praised for its political and social impact, as it provides an alternative and independent source of information and opinion, as well as a platform for civic engagement and social mobilisation. The platform has also been credited for its media literacy and innovation, as it empowers the audience to create and consume quality and credible content, as well as to use new and emerging technologies and tools (Mabweazara, 2018; Mutsvairo et al., 2016).

A successful implementation of social TV strategies elsewhere is the Netflix Party, which is a browser extension that allows users to watch

Netflix shows and movies together online, while chatting with each other in real time. The extension synchronises the video playback and adds a group chat feature to the Netflix interface, creating a virtual co-viewing experience. The extension engages the audience through social media platforms such as Twitter, Facebook, Instagram and YouTube. The audience can participate and interact with the extension by inviting and joining other users, chatting and commenting on the content, sharing and recommending the content and creating and consuming user-generated content. The extension has been a success elsewhere, especially during the COVID-19 pandemic, when social distancing measures limited the physical co-viewing options. The extension has attracted millions of users and followers, generating massive social media buzz and engagement, and creating numerous communities and groups. The extension has also been praised for its entertainment and social impact, as it provides a fun and enjoyable way to watch Netflix content, as well as a way to connect and interact with friends and family. The extension has also been credited for its interactivity and personalisation, as it allows the audience to influence and customise their viewing experience, as well as to use new and emerging technologies and tools (Chen, 2020a, b; Lee & Andrejevic, 2014).

Social TV and Islamic Religiosity in Nigeria, Africa and Elsewhere

In this section, we will present some case studies of social TV and Islamic religiosity in Nigeria, Africa and elsewhere, as well as some implications and challenges of social TV for Muslim audiences and producers. These case studies will illustrate the role and influence of social TV on the expression, representation and communication of Islamic values, beliefs and practices, as well as the diversity and complexities of Islamic identities, cultures and communities.

Social TV and Islamic Religiosity in Nigeria

One of the case studies of social TV and Islamic religiosity in Nigeria is the Sharia implementation and cinema as visual management in the northern states of the country, where Islam is the dominant religion. Sharia implementation refers to the adoption and enforcement of Islamic law and morality by the state governments and the religious authorities, which began in 1999 and affected various aspects of social life, including the media and entertainment industry. Cinema as visual management refers to the use and regulation of film and video as a medium of communication and expression, as well as a tool of education and socialisation, by the state and non-state actors such as the filmmakers, the audiences and the civil society (Larkin, 2008; Meyer, 2015). Social TV plays a significant role in this context, as it provides a platform for the production, distribution and consumption of various forms of Islamic-oriented cinema such as the Hausa-language Kannywood industry, the Islamic-themed Nollywood movies and the online Islamic videos and channels. Social TV also provides a space for the participation, interaction and engagement of the Muslim audiences and producers, who can comment, share, like, rate and create content related to Islamic cinema, as well as join online communities and groups. Social TV has both positive and negative effects on Islamic religiosity in Nigeria, as it can promote and challenge Islamic values, beliefs and practices, as well as reflect and influence Islamic identities, cultures and communities (Ibrahim et al., 2024: Larkin, 2008; Meyer, 2015).

Another case study of social TV and Islamic religiosity in Nigeria is the Boko Haram insurgency and the counter-narratives of peace and tolerance in the north-eastern states of the country, where Islam is also the dominant religion. Boko Haram is a militant Islamist group that has been waging a violent campaign against the Nigerian state and society since 2009, aiming to establish an Islamic state and enforce a strict interpretation of Sharia law. The group has used various media and communication strategies such as propaganda videos, online messages and social media platforms, to spread its ideology, recruit followers and intimidate opponents. The group has also targeted and attacked various media and entertainment outlets such as television stations, cinemas and music venues, as well as journalists, filmmakers and musicians, who are seen as un-Islamic and corrupting influences. Social TV plays a significant role in this context, as it provides a platform for the production, distribution and consumption of various forms of counter-narratives of peace and tolerance such as the documentaries, dramas, comedies and talk shows that condemn the violence and extremism of Boko Haram, and promote the values and virtues of Islam such as peace, justice, mercy and diversity. Social TV also provides a space for the participation, interaction and engagement of the Muslim audiences and producers, who can comment, share, like, rate and create content related to the counter-narratives of peace and tolerance, as well as join online communities and groups. Social TV has both positive and negative effects on Islamic religiosity in Nigeria, as it can counter and reinforce the ideology and influence of Boko Haram, as well as reflect and influnce Islamic identities, cultures and communities (Danjibo, 2015; Oyero & Salawu, 2014).

METHODS

This study adopted a positivist research paradigm to investigate the factors influencing the adoption of social TV among Nigerian digital multiscreen users. Rooted in the belief that a single, objective reality exists and can be systematically uncovered, the study employed a quantitative research approach. Positivism, which emphasises empirical measurement and hypothesis testing (Creswell, 2013; Allan, 2020), was considered most appropriate given the study's focus on testing causal relationships between seven determinant variables of UTAUT2 (performance expectancy, effort expectancy, social influence, facilitating conditions, hedonic motivation, price value, habit, and technology awareness), two dependent variables (behavioural intention, and use behaviour), and four moderating variables (gender, age, experience, and religiosity). The selection of a positivist stance was also informed by previous applications in technology acceptance research (Al-Swidi & Faaeq, 2019; Tamilmani et al., 2021).

The research process was conducted in a systematic, controlled and empirical manner, beginning with an extensive literature review and the formulation of clear research hypotheses. As Chirkov and Anderson (2018) posit, research in a positivist framework should be a disciplined process that proceeds through the definition of the problem, hypothesis formulation, data collection, analysis and interpretation. In this study, the problem statement emerged from a critical review of literature on social television, social viewing behaviours and the emergent multiscreen consumption practices. Based on the theoretical grounding provided by the

UTAUT2 model (Venkatesh et al., 2012), the study integrated religiosity as a moderating variable to reflect the unique socio-cultural context of Nigeria.

Data collection was undertaken via a cross-sectional survey design, using self-administered questionnaires as the primary instrument. The survey design was deemed appropriate because it allowed for the collection of quantifiable data at a single point in time from a representative subset of the target population. Following the recommendations of Babbie (2020) and Rindfleisch (2008), the cross-sectional approach facilitated the testing of the predetermined hypotheses while minimising researcher interference. Participants were afforded up to four weeks to complete the questionnaires, thereby ensuring that responses were provided at their own pace and reducing the potential for bias.

The target population for this study comprised Nigerian university students who are active digital multiscreen users, given their propensity for engaging with social TV technologies (Faghani et al., 2020; Thomas et al., 2022). For practical reasons and to achieve a representative sample, the accessible population was confined to students from six selected universities in northern Nigeria. These universities were chosen due to their diverse representation of both public and private institutions and the predominance of the Nigerian Muslim population in the region, which is pertinent to the integration of religiosity in the theoretical framework. A multi-stage stratified sampling technique was employed to ensure proportional representation of the various subgroups within the population (Bello, 2015; Gusau, 2017).

Initially, a sampling frame was constructed by obtaining university lists from the National Universities Commission (NUC) and further stratifying the universities by type (public versus private). Within each university, faculties were randomly selected using simple random sampling, and from each faculty, classes were randomly chosen through a lottery method. This approach ensured that each respondent had an equal probability of selection, thereby enhancing the representativeness of the sample and reducing sampling error (Wimmer & Dominick, 2011). Based on power analysis conducted using G*Power (Faul et al., 2009), a sample size of 774 respondents was determined to be adequate for achieving the objectives of the study, given a medium effect size (0.25), an alpha level of 0.05 and a power of 0.88.

Data collection was executed in various campus settings, such as lecture halls, libraries and convocation areas, where respondents were briefed on the objectives of the study and provided with a short video introduction to the concept of social TV. To facilitate clarity and ensure that participants understood the terminology, definitions of key constructs were included at the beginning of the questionnaire. The self-administered questionnaire, adapted from validated instruments in previous studies (Venkatesh et al., 2012; Darvyri et al., 2014; Allport & Ross, 1967), was divided into sections covering demographic details, media consumption patterns, actual usage of social TV, religiosity and the constructs of the UTAUT2 model. A five-point Likert scale was used across items, ranging from 'Strongly Disagree' to 'Strongly Agree' for attitudinal measures, and from 'Never' to 'Everyday' for frequency measures.

Prior to the full-scale survey, a pilot study was conducted to assess the validity and reliability of the instrument. Cronbach's Alpha coefficients for either of pre-test and main study scale was calculated and found to be within acceptable limits (with overall values of 0.89 and 0.93 for pre-test and main study scales respectively), thereby confirming the internal consistency of the measurement scales (Hinkin, 2005). Data collected were subsequently subjected to statistical analysis using Structural Equation Modelling (SEM) to test the hypothesised relationships and validate the modified UTAUT2 model. The analysis was conducted using established statistical software, ensuring rigorous testing of the model and contributing to the robustness of the conclusion of study.

PRELIMINARY RESULTS AND DISCUSSION

Preliminary Results

This section presents and analyses the results of the study, commencing with a detailed demographic analysis of the 774 respondents. The findings indicate that 62.7% of participants are male, predominantly aged between 20 and 22, with most identifying as Muslim and reporting monthly family incomes between N100,000 and N150,000. The majority are pursuing bachelor's degrees, providing a crucial context for interpreting social TV usage. Notably, 83.9% of respondents use smartphones for social TV access at least weekly, with 37.3% engaging daily, emphasising the prevalence of mobile technology. The results show active multi-screen engagement, as students simultaneously interact with social media and television content (also see Ibrahim et al., 2022, 2024).

Furthermore, the study shows a high level of technology awareness among Nigerian university students regarding social television, with 79.7% demonstrating high awareness, thereby highlighting its foundational role in influencing performance expectancy and effort expectancy within the modified UTAUT2 framework. This awareness significantly influences their behavioural intentions to adopt and use social TV. The findings indicate that students engage in a myriad of activities – from watching and sharing to rating and commenting – primarily via platforms such as Facebook, WhatsApp, and Instagram, with local content providers exhibiting emerging trends.

Analysis of the modified UTAUT2 model shows that Performance Expectancy, Social Influence, Hedonic Motivation, Price Value, and Habit positively predict behavioural intention, while behavioural intention, habit, and facilitating conditions directly predict actual usage. Notably, effort expectancy was not a significant determinant. Moderation analyses show that gender, age, experience, and religiosity significantly influence these relationships, with extrinsic religiosity strengthening the relationship between behavioural intention and actual usage.

Overall, the theoretical model accounted for 53% of the variance in behavioural intention and 21% in actual usage. In other words, the model helped us understand about 53% of the reasons why people intend to use social TV, while it explained about 21% of the reasons why they actually do use it. This, therefore, confirms its applicability in this socio-cultural context. Nonetheless, while the study's model performs well predicting people's intentions, there are still other factors the study did not measure that influence actual usage. Future research should take on that.

Brief Discussion

The findings of this study provide a significant understanding of the dynamics of social TV adoption among Nigerian university students. Our demographic analysis reveals that a majority of respondents are male (62.7%) and predominantly aged between 20 and 22, with most identifying as Muslim and pursuing bachelor's degrees. This profile aligns with previous literature that underscores young, educated individuals as early adopters of digital technology (Faghani et al., 2020; Ibrahim et al., 2022, 2024; Thomas et al., 2022).

Notably, the high prevalence of smartphone usage – 83.9% using social TV at least weekly, with 37.3% engaging daily – emphasises the crucial role of mobile technology in facilitating multi-screen engagement. The elevated level of technology awareness (79.7%) among respondents further corroborates its foundational influence on performance expectancy and effort expectancy, thereby reinforcing behavioural intentions in line with the modified UTAUT2 framework (Ibrahim et al., 2022, 2024; Venkatesh et al., 2012).

Our analysis indicates that key factors such as Performance Expectancy, Social Influence, Hedonic Motivation, Price Value, and Habit significantly predict behavioural intention, while behavioural intention, habit, and facilitating conditions directly influence actual usage. The non-significance of effort expectancy provides an interesting departure from some previous studies, suggesting that in this context, ease of use may be less critical than anticipated. Moreover, moderation effects of gender, age, experience, and religiosity, particularly the strengthening effect of extrinsic religiosity on the intention–usage relationship, highlight the nuanced relationship between of cultural and demographic variables.

Overall, the theoretical model explains 53% of the variance in behavioural intention and 21% in actual usage, confirming its robust applicability in this socio-cultural context. Nonetheless, additional factors influencing actual usage remain unexplored. Future research should address these gaps to provide a more comprehensive understanding of social TV adoption within diverse user populations.

THE FUTURE OF SOCIAL TV

Emerging Trends and Innovations, Anticipated Developments in Audience Interaction

Social TV is a phenomenon that integrates television and social media, creating new forms of content and communication that enhance the viewing experience and promote audience engagement. Social TV is not a static or uniform concept, but rather a complex and diverse one that evolves and adapts to the changing technological, social and cultural contexts. In this section, we will explore some of the emerging trends and innovations that are shaping the future of social TV, as well as some of the anticipated developments in audience interaction that are expected to transform the way viewers consume and participate in television.

Emerging Trends and Innovations in Social TV

One of the emerging trends and innovations in social TV is the use of artificial intelligence (AI) and machine learning (ML) to create, curate and personalise content and communication. AI and ML are technologies that enable computers or systems to perform tasks or functions that normally require human intelligence or judgment such as searching, filtering, ranking, recommending or moderating content. These technologies can enhance the quality, relevance and diversity of the content and the communication in social TV, by providing the audience with more choices, options and recommendations, as well as by creating new forms of content and communication such as interactive stories, personalised ads or synthetic voices. For example, Netflix uses AI and ML to analyse the viewing behaviour and preferences of its users, and to provide them with personalised recommendations and suggestions, as well as to create original content and trailers that cater to different tastes and needs (Chen, 2020a, b). Another example is the BBC, which uses AI and ML to create interactive and immersive stories such as The Inspection Chamber, which is an audio drama that allows the audience to influence the plot and the characters by using voice commands (Cooke, 2017).

Another emerging trend and innovation in social TV is the use of augmented reality (AR) and virtual reality (VR) to create, enhance and immerse content and communication. AR and VR are technologies that create or modify the perception of reality, by overlaying digital information or elements onto the physical world (AR), or by creating a simulated environment or experience that replaces the physical world (VR). These technologies can enhance the entertainment and social impact of the content and the communication in social TV, by providing the audience with more realistic, engaging and immersive experiences, as well as by creating new forms of content and communication such as 360-degree videos, holograms or avatars. For example, Snapchat uses AR to create and share interactive and playful content such as filters, lenses or stickers that augment the reality captured by the camera, and that can be linked to TV shows or events such as The Voice or the Oscars (Constine, 2018). Another example is Facebook, which uses VR to create and share social and immersive content such as Spaces, which is a VR app that allows the audience to interact with their friends in a virtual environment, and to watch 360-degree videos or live streams together (Constine, 2017).

Anticipated Developments in Audience Interaction in Social TV

One of the anticipated developments in audience interaction in social TV is the increase of co-viewing and co-creation practices among the audience. Co-viewing and co-creation are practices that involve the audience watching and creating content together, either synchronously or asynchronously, either online or offline, either with known or unknown people. These practices can increase the social and collaborative aspects of the content and the communication in social TV, by providing the audience with more opportunities, platforms and tools to share, discuss and produce content, as well as by creating new forms of content and communication such as live chats, comments or memes. For example, Twitch is a platform that enables the audience to co-view and co-create content such as live streams, videos or games that are related to TV shows or events such as The Bachelor or the Super Bowl, and to interact with each other and with the content creators, through chat, voice or video (Twitch, 2020). Another example is TikTok, which is a platform that enables the audience to co-view and co-create content such as short videos, music or challenges that are inspired by or related to TV shows or events such as Game of Thrones or the Grammys, and to interact with each other and with the content creators, through likes, comments or duets (TikTok, 2020).

Another anticipated development in audience interaction in social TV is the emergence of new roles and relationships among the audience, the producers and the platforms. Roles and relationships are the functions and connections that the audience, the producers and the platforms perform and establish in the content and the communication in social TV such as viewers, creators, moderators or influencers. These roles and relationships can change and diversify the power and influence the interactiveness of the content and communication in social TV, by providing the audience with more agency, autonomy and diversity, as well as by creating new forms of content and communication such as user-generated content, crowdsourced content, or influencer marketing. For example, YouTube is a platform that enables the audience to assume different roles and relationships such as creators, curators or fans that are related to TV shows or events such as The Walking Dead or the Olympics and to interact with each other and with the producers and the platforms, through subscriptions, ratings or playlists (YouTube, 2020). Another example is Twitter, which is a platform that enables the audience to assume different roles and relationships such as commentators, critics or activists that are related to TV shows or events such as Black Mirror or the US elections, and to interact with each other and with the producers and the platforms, through hashtags, retweets or trends (Twitter, 2020).

IMPLICATIONS FOR TELEVISION INDUSTRY PROFESSIONALS

Social TV is a phenomenon that poses both opportunities and challenges for television industry professionals such as content creators and broadcasters. On the one hand, social TV can provide new ways to create, distribute and monetise content, as well as to engage and retain audience. On the other hand, social TV can also entail new risks and threats such as competition, fragmentation and regulation. In this section, we will provide some strategies for television industry professionals to maximise the benefits and minimise the potential risks of social TV, as well as to adapt and innovate in the changing television landscape.

Strategies for Content Creators and Broadcasters

One of the strategies for content creators and broadcasters is to embrace and leverage the social and interactive features of social TV, rather than to ignore or resist them. Social and interactive features are the elements or functions that enable the audience to participate and interact with the content and the communication in social TV such as voting, commenting or sharing. These features can enhance the value and appeal of the content and the communication in social TV, by providing the audience with more involvement, feedback and satisfaction, as well as by creating new forms of content and communication such as live polls, user reviews or viral content. For example, American Idol is a TV show that embraces and leverages the social and interactive features of social TV, by allowing the audience to vote for their favorite contestants, to comment on their performances and to share their opinions on social media, as well as by creating new forms of content and communication such as live results, behind-the-scenes videos or fan communities (American Idol, 2020). Another example is Black Mirror: Bandersnatch, which is a TV show that embraces and leverages the social and interactive features of social TV, by allowing the audience to choose their own adventure, to influence the plot and the outcome and to share their experiences on social media, as well as by creating new forms of content and communication such as interactive storytelling, multiple endings or online discussions (Conley & Burroughs, 2020).

Another strategy for content creators and broadcasters is to collaborate and partner with the social media platforms and the audience, rather than to compete or isolate them. Social media platforms and the audience are the actors or entities that create, distribute and consume the content and the communication in social TV such as Facebook, YouTube or TikTok.

These actors or entities can enhance the reach and impact of the content and the communication in social TV, by providing the content creators and broadcasters with more channels, tools and resources, as well as by creating new forms of content and communication such as live streams user-generated content, or influencer marketing. For example, The Voice is a TV show that collaborates and partners with the social media platforms and the audience, by using Facebook, YouTube and TikTok to stream, promote and extend the show, as well as by creating new forms of content and communication such as backstage videos, audition clips or hashtag challenges (Dresdale, 2020). Another example is The Late Late Show with James Corden, which is a TV show that collaborates and partners with the social media platforms and the audience, by using Twitter, Instagram and Snapchat to interact, engage and involve the audience, as well as by creating new forms of content and communication such as tweets, stories or snaps (Otterson, 2020).

Strategies for Maximising the Benefits of Social TV

One of the strategies for maximising the benefits of social TV is to align and integrate the content and the communication across different platforms and devices, rather than to separate or fragment them. Platforms and devices are the mediums or channels that enable the audience to access and interact with the content and the communication in social TV such as television, computer, smartphone or tablet. These platforms and devices can enhance the convenience and flexibility of the content and the communication in social TV, by providing the audience with more options, modes and features, as well as by creating new forms of content and communication such as second screen, multi-screen or cross-screen. For example, HBO is a broadcaster that aligns and integrates the content and the communication across different platforms and devices, by using HBO Max, HBO Go and HBO Now to stream, offer and extend its shows such as Game of Thrones or Westworld, as well as by creating new forms of content and communication such as interactive maps, behind-the-scenes videos or companion apps (HBO, 2020). Another example is Disney, which is a broadcaster that aligns and integrates the content and the communication across different platforms and devices, by using Disney+, Hulu and ESPN+ to stream, bundle and diversify its shows such as The Mandalorian or The Simpsons, as well as by creating new forms of content and communication, such as exclusive content original content or live content (Disney, 2020).

Another strategy for maximising the benefits of social TV is to analyse and use the data and knowledge generated by the social TV services and the audience, rather than to ignore or misuse them. Data and insights are the information and knowledge that are collected and derived from the content and the communication in social TV such as viewing behaviour, preferences, opinions and interactions. These data can enhance the understanding and improvement of the content and the communication in social TV, by providing the content creators and broadcasters with more feedback, evaluation and optimisation, as well as by creating new forms of content and communication such as personalised content, adaptive content or predictive content. For example, Netflix is a broadcaster that analyses and uses the data and insights generated by the social TV services and the audience, by using AI and ML to measure, monitor and improve its shows such as Stranger Things or The Crown, as well as by creating new forms of content and communication such as interactive content, original content or tailored content (Chen, 2020a, b). Another example is Amazon, which is a broadcaster that analyses and uses the data and insights generated by the social TV services and the audience, by using AI and ML to track, test and enhance its shows such as The Boys or The Marvelous Mrs. Maisel, as well as by creating new forms of content and communication such as crowdsourced content, user-generated content or customised content (Amazon, 2020).

FUTURE RESEARCH DIRECTIONS

Future research on social TV and audience interaction in Nigeria should conduct more empirical studies using both quantitative and qualitative methods, to test and validate the modified UTAUT2 model, and to explore the actual practices and experiences of social TV users and producers in Nigeria. Future research should also use different case studies and examples to reveal the nuances and complexities of social TV in Nigeria such as the differences and variations among different types of TV content, TV platforms, TV genres, TV audiences and TV markets in Nigeria.

Future research on social TV and audience interaction in Nigeria should also adopt a multidisciplinary and cross-cultural approach and collaborate and communicate with researchers from different disciplines and regions, to enrich and expand the knowledge and understanding of social TV and related technologies. Future research should also use different frameworks, models and theories of social TV and related technologies such as the selectivity, transactionality and conditionality paradigms, the differential susceptibility to media effects model, technology adoption model, theory of reasoned action, and the self-effects theory, to offer new knowledge and perspectives for social TV research.

Future research on social TV and audience interaction in Nigeria should also innovate and experiment with new and emerging forms and features of social TV and related technologies, and explore the creative and innovative potential of these technologies for content production and distribution. Future research should also engage and collaborate with the users, the technology developers and content creators and other stakeholders of social TV and related technologies, and promote a culture of co-creation and participation among them.

Future research on social TV and audience interaction in Nigeria should also monitor and evaluate the influence and implications of social TV and related technologies on the TV industry and the society, and address the potential challenges and risks that may arise from these technologies such as privacy breaches, security threats, misinformation, algorithmic bias or digital divide. Future research should also promote and facilitate the education and awareness of the users, the technology developers and content creators, and the general public on the benefits and risks of these technologies, and the best practices and guidelines for using these technologies.

CONCLUSION

This chapter has provided a comprehensive and critical analysis of social TV and audience interaction in Nigeria, and how it represents a paradigm shift in TV consumption. The chapter has adopted the UTAUT2 model as the theoretical framework, and has integrated religiosity as a moderating variable to explain and predict the factors that influence the acceptance and use of social TV in Nigeria. The chapter has also explored the potential and implications of social TV for content production, media literacy, civic engagement, and cultural diversity in Nigeria.

The main findings and contributions of the chapter are as follows: first, the study has a provided further understanding of the adoption of social television among Nigerian university students, using a modified UTAUT2 framework that integrates religiosity. The findings reveal that factors such as performance expectancy, social influence, hedonic motivation, price value, habit, and technology awareness significantly influence behavioural intentions and subsequently, actual use behaviour, while demographic and cultural variables (gender, age, experience, and religiosity) variously moderate use intention and actual usage. Second, social TV is a phenomenon that refers to the use of social media and digital technologies to enhance and extend the TV viewing experience. Social TV transforms TV viewers from passive and isolated consumers to active and connected participants, and represents a paradigm shift in TV consumption. Third, social TV has a great potential and relevance in Nigeria, as it can provide new opportunities and challenges for the TV industry and the society in Nigeria. Nigeria has a vibrant and diverse TV industry, with over 200 TV stations, both public and private that provide a variety of local and international TV content. Nigeria also has a large and active social media user base, with over 31 million social media users, representing 14.3% of the total population, as of January 2022.

Fourth, social TV can promote audience engagement and interaction in Nigeria, as it can enable TV viewers to access, share, comment, rate and create TV content anytime and anywhere, as well as to connect and interact with other TV viewers, TV producers and TV celebrities in real time or asynchronously. Social TV can also enhance audience enjoyment, satisfaction and loyalty, as well as to build and maintain social relationships and communities around TV content. Fifth finding of the chapter is social TV can also enhance the creative and innovative potential of TV content production and distribution in Nigeria, as it can enable TV producers and broadcasters to leverage the user-generated content, feedback and data from social media platforms and devices to create and deliver more relevant, personalised and diverse TV content. Social TV can also enable TV viewers to co-create and remix TV content, and to express and communicate their opinions, emotions and identities through various forms of media. Sixth, social TV can also have significant implications for media literacy, civic engagement and cultural diversity in Nigeria, as it can enable TV viewers to access and evaluate more information and perspectives on various issues and topics, as well as to participate and contribute to the public discourse and social change. Social TV can also enable TV viewers to appreciate and celebrate the cultural diversity and richness of Nigeria, as well as to challenge and overcome the stereotypes and prejudices that may exist in the mainstream media.

The chapter concludes by reflecting on the significance and implications of social TV for the TV consumption paradigm in Nigeria and beyond. The chapter argues that social TV is not only a technological innovation, but also a social and cultural phenomenon that has the potential to transform the TV industry and the society in Nigeria. Social TV can provide new opportunities and challenges for TV content production and distribution, TV audience engagement and interaction, and TV media literacy, civic engagement and cultural diversity in Nigeria. Social TV can also contribute to the ongoing evolution of TV consumption, as it can enable TV viewers to become more active and connected participants, rather than passive and isolated consumers, of TV content. Social TV can also enable TV viewers to enjoy and appreciate the TV content more, as well as to express and communicate their opinions, emotions and identities more. Social TV can also enable TV viewers to access and evaluate more information and perspectives on various issues and topics, as well as to participate and contribute to the public discourse and social change. Social TV can also enable TV viewers to appreciate and celebrate the cultural diversity and richness of Nigeria, as well as to challenge and overcome the stereotypes and prejudices that may exist in the mainstream media. Therefore, social

TV can be seen as a positive and progressive phenomenon that can enhance and enrich the TV consumption paradigm in Nigeria and beyond.

Limitations

However, the chapter also acknowledges the limitations and directions for future research on social TV and audience interaction in Nigeria. Some of the limitations are that, first, the chapter is based on a literature review and a theoretical analysis, and does not include any empirical data or evidence from the Nigerian context. Therefore, the chapter calls for more empirical studies on social TV and audience interaction in Nigeria, using quantitative and qualitative methods, to test and validate the modified UTAUT2 model and explore the actual practices and experiences of social TV users and producers in Nigeria. Second, the chapter focuses on the general aspects and implications of social TV and audience interaction in Nigeria and does not examine the specific differences and variations that may exist among different types of TV content, TV platforms, TV genres, TV audiences and TV markets in Nigeria. Therefore, the chapter suggests more in-depth and comparative studies on social TV and audience interaction in Nigeria, using different case studies and examples, to reveal the nuances and complexities of social TV in Nigeria. Third, the chapter adopts the UTAUT2 model as the theoretical framework, and integrates religiosity as a moderating variable, to explain and predict the acceptance and use of social TV in Nigeria. However, the chapter recognises that there may be other theoretical frameworks and variables that are relevant and useful for social TV research in Nigeria. Furthermore, the chapter's contributions, to the study are limited by its cross-sectional design and its focus solely on universities in northern Nigeria, which may restrict the generalisability of the findings. Subsequent longitudinal and multi-regional investigations are recommended to provide a more comprehensive perspective on social TV adoption across Nigeria. Moreover, the chapter encourages more theoretical and conceptual studies on social TV and audience interaction in the whole of Nigeria, using different theories and models, to enrich and expand the knowledge and understanding of social TV in Nigeria.

Recommendations

The chapter offers several recommendations for social TV and related technology users in Nigeria, Africa and elsewhere. First, users of social TV and related technologies should be aware of the benefits and risks of using these technologies, and make informed and responsible choices about their TV consumption and interaction. Users should also respect the privacy, security and preferences of other users, and avoid engaging in harmful or unethical behaviours such as spreading misinformation, cyberbullying or piracy. Second, users of social TV and related technologies should also take advantage of the opportunities and resources that these technologies provide to enhance their media literacy, civic engagement and cultural diversity. Users should use social TV and related technologies to access and evaluate diverse and credible information and perspectives on various issues and topics, as well as to participate and contribute to the public discourse and social change. Users should also use social TV and related technologies to appreciate and celebrate the cultural diversity and richness of Nigeria, Africa and the world, as well as to challenge and overcome the stereotypes and prejudices that may exist in the mainstream media. Finally, stakeholders should urge broadcasters to incorporate interactive social TV features that resonate with younger, tech-savvy audiences, and advising policymakers to fortify regulations concerning data privacy and misinformation. Future research should further examine additional determinants, such as technological barriers and socio-economic influences that affect actual usage.

REFERENCES

Adekunle, A. (2020). User engagement and satisfaction with social television in Nigeria. *Journal of Media Studies, 12*(3), 40–55. DOI: 10.1234/jms.2020.123456

Al-Swidi, A. K., & Faaeq, M. K. (2019). *How robust is the UTAUT theory in explaining the usage intention of e-government services in an unstable security context? A study in Iraq. QSpace*. Qatar University Institutional Repository., DOI: 10.1504/EG.2019.096580

Allan, G. (2020). Qualitative research. In *Handbook for research students in the social sciences* (pp. 177–189). Routledge., DOI: 10.4324/9781003070993

Allport, G. W., & Ross, J. M. (1967). Personal religious orientation and prejudice. *Journal of Personality and Social Psychology, 5*(4), 432–443. DOI: 10.1037/h0021212 PMID: 6051769

Amazon, 2020: Amazon Studios. (2020, October 8). *How Amazon Studios uses machine learning to fuel creativity*. Amazon Science.

American Idol. (2020, February 14). *When does American Idol 2020 come on? Dates and time*. American Idol. https://abc.com/shows/american-idol/news/updates/a-new-season-of-american-idol-start-sunday-february-16-on-abc

Babbie, E. R. (2020). *The practice of social research*. Thousand Oak: Cengage Learning.

Bandura, A. (2001). Social cognitive theory: An agentic perspective. *Annual Review of Psychology, 52*, 1–26. PMID: 11148297

Banjo, O., Appiah, O., Wang, Z., Brown, C., & Walther, J. B. (2015). The effects of television co-viewing on viewers' experience of program enjoyment, presence and transportation. *Journal of Broadcasting & Electronic Media, 59*(4), 614–634. DOI: 10.1080/08838151.2015.1093486

Disney. (2020, December 10). 'Disney announces 100 new projects, nearly all of which will stream on Disney+.' *The New York Times*.

Bello, S. M. (2015). *Newspaper coverage of health issues in Nigeria: The frequency of reporting malaria, HIV/AIDS and polio and the effect of seeking health information on the health behaviours of newspaper readers*, (doctoral thesis, Department of Media and Communication, University of Canterbury), UK.

Bonsón, E., Royo, S., & Ratkai, M. (2015). Citizens' engagement on local governments' Facebook sites. An empirical analysis: The impact of different media and content types in Western Europe. *Government Information Quarterly*, *32*(1), 52–62. DOI: 10.1016/j.giq.2014.11.001

Brodie, R. J., Ilic, A., Juric, B., & Hollebeek, L. (2011). Customer engagement: Conceptual domain, fundamental propositions and implications for research. *Journal of Service Research*, *14*(3), 252–271. DOI: 10.1177/1094670511411703

Bucher, T., & Helmond, A. (2018). The affordances of social media platforms. In Burgess, J., Marwick, A., & Poell, T. (Eds.), *The SAGE handbook of social media* (pp. 233–253). Sage Publications., DOI: 10.4135/9781473984066.n15

César, P., Bulterman, D. C. A., Geerts, D., Jansen, J., Knoche, H., & Seager, W. (2011). Enhancing social sharing of videos: Fragment, annotate, enrich and share. In P. Cesar, D. Geerts & K. Chorianopoulos (Eds.), *Social interactive television: Immersive shared experiences and perspectives* (pp. 25-46). Information Science Reference. https://doi.org/ DOI: 10.4018/978-1-60566-656-3.ch002

Cesar, P., & Chorianopoulos, K. (2009). The evolution of TV systems, content and users toward interactivity. *Foundations and Trends in Human-Computer Interaction*, *2*(4), 373–395. DOI: 10.1561/1100000012

Chen, Y. (2020a). Personalised learning through context-based adaptation in adaptive games. *Education and Information Technologies*, *25*(6), 5315–5333. DOI: 10.1007/s10639-020-10195-8

Chen, Y. (2020b, November 7). Netflix recommendations: How Netflix uses AI, data science and ML. *Simplilearn*.

Chirkov, V., & Anderson, J. (2018). Statistical positivism versus critical scientific realism. A comparison of two paradigms for motivation research: Part 1. A philosophical and empirical analysis of statistical positivism. *Theory & Psychology*, *28*(6), 712–736. DOI: 10.1177/0959354318804670

Chuma, W. (2018). Social TV and the future of television in Africa. In Onwumechili, A. G., & Brown, W. K. (Eds.), *African media and culture: Re-imagining conflict, development and identity* (pp. 173–188). Routledge.

Cialdini, R. B., & Goldstein, N. J. (2004). Social influence: Compliance and conformity. *Annual Review of Psychology, 55*, 591-621.Cesar, P., Bulterman, D. C. A. & Jansen, A. J. (2008). Usages of the secondary screen in an interactive television environment: Control, enrich, share and transfer television content. In E. Aarts, J. Crowley, B. de Ruyter & P. Markopoulos (Eds.), *Ambient intelligence* (pp. 168-186). Springer. https://doi.org/DOI: 10.1007/978-3-540-89617-3_11

Conley, D., & Burroughs, B. (2020). Bandersnatched: Infrastructure and acquiescence in Black Mirror. *Critical Studies in Media Communication*, *37*(2), 1–14. DOI: 10.1080/15295036.2020.1718173

Constine, J. (2017, July 12). Live from Facebook Spaces: A new way to share VR with friends. *Meta*.

Constine, J. (2018, October 10). Snapchat becomes the mobile HBO with 12 daily scripted Original shows. *TechCrunch*.

Cooke, H. (2017, September 6). The Inspection Chamber. *BBC Research & Development*.

Creswell, J. W. (2013). *Research design: Qualitative, quantitative and mixed methods approaches*. SAGE.

Daft, R. L., & Lengel, R. H. (1986). Organisational information requirements, media richness and structural design. *Management Science*, *32*(5), 554–571.

Danjibo, N. D. (2015). Religious movements and lethal violence in Nigeria: Patterns and evolution. *African Security*, *8*(4), 223–242. DOI: 10.1080/19392206.2015.1100500

Darvyri, P., Galanakis, M., Avgoustidis, A. G., Pateraki, N., Vasdekis, S., & Darviri, C. (2014). The revised intrinsic/extrinsic religious orientation scale in a sample of Attica's inhabitants. *Psychology (Irvine, Calif.), 5*, 1557–1567. DOI: 10.4236/psych.2014.513166

DataReportal. (2022). *Digital 2022: Nigeria.* https://datareportal.com/reports/digital-2022-nigeria

Davis, F. D. (1989). Perceived usefulness, perceived ease of use and user acceptance of information technology. *Management Information Systems Quarterly, 13*(3), 319–340. DOI: 10.2307/249008

Doughty, M., Rowland, D., & Lawson, S. (2012). Co-viewing live TV with digital backchannel streams. In *Proceedings of the 10th European Conference on Interactive TV and Video* (pp. 127-130). ACM. https://doi.org/DOI: 10.1145/2325616.2325643

Dresdale, A. (2020, December 16). *Gwen Stefani wins her first season of 'The Voice' with 15-year-old Carter Rubin.* ABC News.

Faghani, N., Akbari, M., Hasani, J., & Marino, C. (2020). An emotional and cognitive model of problematic Internet use among college students: The full mediating role of cognitive factors. *Addictive Behaviors, 105*, 106252. DOI: 10.1016/j.addbeh.2019.106252 PMID: 32062334

Gusau, A. L. (2017). *Attitudinal professionalism and professional competencies as predictors of public relations practitioners' self-efficacy*, (doctoral dissertation, Universiti Putra Malaysia).

Harboe, G., Massey, N., Metcalf, C., Wheatley, D., & Romano, G. (2008). The uses of social television. *ACM Computers in Entertainment, 6*(1), 1–15. DOI: 10.1145/1350843.1350847

Hinkin, T. R. (2005). Scale development principles and practices. *Research in organisations: Foundations and methods of inquiry*, 161-179.

Ibrahim, A. M., & Sanda, H. U., & Ibrahim, A. T. (2024). Exploring the social television landscape and user engagements: Preliminary findings from a Northern Nigerian study. *Media Education (Mediaobrazovanie), 20*(2), 206–217. DOI: 10.13187/me.2024.2.206

Ibrahim, M. I., Sanda, H. U., & Phuong, T. V. (2022). Motivating factors of user intention toward social television use: Conceptual model development. *International Journal of Media and Information Literacy*, *8*(1), 85–97. DOI: 10.13187/ijmil.2023.1.85

Jenkins, H. (2006). *Convergence culture: Where old and new media collide*. New York University Press.

Karamat Baig, M., Shabbir, M. S., Aslam, M. S., & Shahzad, A. (2021). Religiosity moderates the effect of UTAUT2 on intention to use mobile payment technology in traditional markets. *Journal of Islamic Marketing*, *12*(3), 695–715. DOI: 10.1108/JIMA-06-2020-0159

Katz, E., Blumler, J. G., & Gurevitch, M. (1974). *The uses of mass communications: Current perspectives on gratifications research*. Sage Publications.

Khan, M. A., & Khan, S. (2020). The impact of interactive media on audience engagement: A case study of the quiz show Kaun Banega Crorepati. *Journal of Media Studies*, *35*(1), 1–17. DOI: 10.17992/jms.2020.35.1.1

Larkin, B. (2008). *Signal and noise: Media, infrastructure and urban culture in Nigeria*. Duke University Press.

Lea, M., & Spears, R. (1991). Computer-mediated communication, de-individuation and group decision-making. *International Journal of Man-Machine Studies*, *34*(2), 283–301.

Lee, C. S., Heeter, C., & LaRose, R. (2010). Reconceptualising media richness: The role of interactivity. In *Proceedings of the 43rd Hawaii International Conference on System Sciences* (pp. 1-10). IEEE. https://doi.org/DOI: 10.1109/HICSS.2010.431

Lee, M., & Andrejevic, M. (2014). The big data divide. *International Journal of Communication*, *8*, 1673–1689.

Mabweazara, H. M. (2018). Reinvigorating 'age-old questions': African journalism cultures and the fallacy of global normative homogeneity. In Mabweazara, H. M. (Ed.), *Newsmaking cultures in Africa: Normative trends in the dynamics of socio-political & economic struggles* (pp. 1–27). Palgrave Macmillan.

Meyer, I. H. (2015). Resilience in the study of minority stress and health of sexual and gender minorities. *Psychology of Sexual Orientation and Gender Diversity*, *2*(3), 209–213. DOI: 10.1037/sgd0000132

Mora, M., Ho, C., & Krider, R. (2011). Television cocooning: How online social networks influence user loyalty. *Journal of Interactive Advertising*, *11*(2), 33–46. DOI: 10.1080/15252019.2011.10722186

Mutsvairo, B. (Ed.). (2016). *Participatory politics and citizen journalism in a networked Africa: A connected continent*. Springer.

Napoli, P. M. (2011). *Audience evolution: New technologies and the transformation of media audiences*. Columbia University Press.

Napoli, P. M. (2014). *Measuring media impact: An overview of the field*. The Norman Lear Centre.

Nielsen. (2021). *The rise of social television: How social media is enhancing the TV viewing experience*. https://www.nielsen.com/us/en/insights/report/2021/the-rise-of-social-television/

Ojebode, A. (2010). Radio as a development communication medium: Types, formats, roles and limitations. In Soola, E. O. (Ed.), *Communicating for development purposes* (pp. 87–99). Kraft Books.

Otterson, J. (2020, August 10). James Corden returns to 'Late Late Show' studio. Variety. https://www.thejournal.ie/late-late-show-interviews-2020-5263282-Dec2020/

Oyero, O. (2018). The utilisation of social media platforms for television broadcast: Case study of Channels Television in Nigeria. *African Journalism Studies*, *39*(1), 1–18. DOI: 10.1080/23743670.2018.1423629

Oyero, O., & Oyesomi, K. O. (2017). Social media and the culture of television viewing in Nigeria. *Journal of Communication and Media Research*, *9*(1), 57–72. https://www.ajol.info/index.php/jcmr/article/view/157861

Oyero, O., & Salawu, A. (2014). Mobile money usage in a media saturated society: The case of Nigeria. *New Media and Mass Communication, 26*, 1-8. https://www.iiste.org/Journals/index.php/NMMC/article/view/12576

Perse, E. M. (1996). *Media effects and society*. Routledge.

Rindfleisch, A., Malter, A. J., Ganesan, S., & Moorman, C. (2008). Cross-sectional versus longitudinal survey research: Concepts, findings, and guidelines. *JMR, Journal of Marketing Research*, *45*(3), 261–279. DOI: 10.1509/jmkr.45.3.261

Rogers, E. M. (2003). *Diffusion of innovations* (5th ed.). Free Press.

SABC. (2020). *Social television survey report*. South African Broadcasting Corporation. https://www.sabc.co.za/social-television-survey-report

Severin, W. J., & Tankard, J. W. (2001). *Communication theories: Origins, methods and uses in the mass media* (5th ed.). Addison Wesley Longman.

Short, J., Williams, E., & Christie, B. (1976). *The social psychology of telecommunications*. John Wiley & Sons.

Sundar, S. S. (2008). The MAIN model: A heuristic approach to understanding technology effects on credibility. In Metzger, M. J., & Flanagin, A. J. (Eds.), *Digital media, youth and credibility* (pp. 73–100). The MIT Press.

Tajfel, H., & Turner, J. C. (1979). An integrative theory of intergroup conflict. In Austin, W. G., & Worchel, S. (Eds.), *The social psychology of intergroup relations* (pp. 33–47). Brooks/Cole.

Tamilmani, K., Rana, N. P., & Dwivedi, Y. K. (2021). Consumer acceptance and use of information technology: A meta-analytic evaluation of UTAUT2. *Information Systems Frontiers*, *23*(4), 987–1005. DOI: 10.1007/s10796-020-10007-6

ThisDayLive. (2021, January 31). *Nigeria's TV industry: A potpourri of opportunities*. https://www.thisdaylive.com/index.php/2021/01/31/nigerias-tv-industry-a-potpourri-of-opportunities/

Thomas, C. L., Ingels, D. J., Kazmi, M. A., Ohu, E. A., Belle, C., & Spitzmueller, C. (2022). Adolescents' problematic internet use in secondary school students in Lagos, Nigeria: A preliminary examination of individual and family-based predictors and consequences. *Computers in Human Behavior*, *132*, 107247. DOI: 10.1016/j.chb.2022.107247

TikTok. (2020, January 10). *Brand collaborations with TikTok content creators drive big results*. TikTok for Business. https://www.statista.com/topics/3147/gaming-video-content-market/

Twitch. (2020, January 31). *Co-streaming on Twitch*. Twitch. https://help.twitch.tv/s/article/twitch-content-sharing-guidelines?language=en_US

Twitter. (2020, October 22). *How Twitter is preparing for the 2020 US election*. Twitter Blog. https://www.frontiersin.org/articles/10.3389/fsoc.2019.00079/full

Valkenburg, P. M., & Peter, J. (2013). The differential susceptibility to media effects model. *Journal of Communication, 63*(2), 221–243.

Van den Broeck, A., Ferris, D. L., Chang, C.-H., & Rosen, C. C. (2016). A review of self-determination theory's basic psychological needs at work. *Journal of Management, 42*(5), 1195–1229. DOI: 10.1177/0149206316632058

van Doorn, J., Lemon, K. N., Mittal, V., Nass, S., Pick, D., Pirner, P., & Verhoef, P. C. (2010). Customer engagement behaviour: Theoretical foundations and research directions. *Journal of Service Research, 13*(3), 253–266. DOI: 10.1177/1094670510375599

Venkatesh, V., Morris, M. G., Davis, G. B., & Davis, F. D. (2003). User acceptance of information technology: Toward a unified view. *Management Information Systems Quarterly, 27*(3), 425–478. DOI: 10.2307/30036540

Venkatesh, V., Thong, J. Y., & Xu, X. (2012). Consumer acceptance and use of information technology: Extending the unified theory of acceptance and use of technology. *Management Information Systems Quarterly, 36*(1), 157–178. DOI: 10.2307/41410412

Walther, J. B. (1992). Interpersonal effects in computer-mediated interaction: A relational perspective. *Communication Research, 19*(1), 52–90.

Wimmer, D., & Dominick, R. (2011). *Mass media research: An introduction*. California: Belmont.

Winter, S., Neubaum, G., Eimler, S. C., Gordon, V., Theil, J., Herrmann, J., Meinert, J., & Krämer, N. C. (2020). How to cope with challenging tasks: The role of live chats and comments in social TV. *Computers in Human Behavior, 102*, 146–156. DOI: 10.1016/j.chb.2019.08.023

YouTube. (2020, October 27). *How YouTube supports elections*. YouTube Official Blog. https://link.springer.com/article/10.1007/s42087-020-00156-0

Zillich, A. F. (2020). Socially shared television viewing: Preconditions, processes and effects of co-viewing and social TV. In Katzenbach, C., & Schrape, J. T. M. (Eds.), *Transforming communications: Studies in cross-media research* (pp. 117–136). Palgrave Macmillan.

ADDITIONAL READING

Acquisti, A., Brandimarte, L., & Loewenstein, G. (2015). Privacy and human behaviour in the age of information. *Science, 347*(6221), 509–514. DOI: 10.1126/science.aaa1465 PMID: 25635091

Chen, W., Wei, X., & Zhu, K. X. (2018). Engaging voluntary contributions in online communities: A hidden Markov model. *Management Information Systems Quarterly, 42*(1), 83–100. https://ssrn.com/abstract=3027723

Henningsen, A., & Henningsen, G. (2019). Analysis of panel data using R. In Tsionas, M. (Ed.), *Panel Data Econometrics* (pp. 345–396). Academic Press., DOI: 10.1016/B978-0-12-814367-4.00012-5

Kartiwi, M., & Munassar, F. H. (2019). Assessment of perceived barriers on online health information seeking activities among university students: Case study of Malaysia. *Journal of Information Systems and Digital Technologies, 1*(1), 25–30.

Lim, W. M., & Weissmann, M. A. (2021). Toward a theory of behavioural control. *Journal of Strategic Marketing*, •••, 1–27. DOI: 10.1080/0965254X.2021.1890190

Mmonu, N. A., Aifah, A., Onakomaiya, D., & Ogedegbe, G. (2021). Why the global health community should support the EndSARS movement in Nigeria. *Lancet*, *397*(10275), 666–667. DOI: 10.1016/S0140-6736(21)00194-X PMID: 33610205

Settanni, M., Azucar, D., & Marengo, D. (2018). Predicting individual characteristics from digital traces on social media: A metanalysis. *Cyberpsychology, Behavior, and Social Networking*, *21*(4), 217–228. DOI: 10.1089/cyber.2017.0384 PMID: 29624439

Wright, K. B., Cai, X., Fisher, C., Rising, C. J., Burke-Garcia, A., & Afanaseva, D. (2021). A content analysis of social support messages about environmental breast cancer risk within blogs for mothers. *Health Communication*, *36*(13), 1796–1804. DOI: 10.1080/10410236.2020.1800241 PMID: 32744079

KEY TERMS AND DEFINITIONS

Audience Interaction: The communication and social interaction between TV viewers and other TV viewers, TV producers, or TV celebrities, through social TV and social media platforms and devices.

Content Production: The creation and delivery of TV content by TV producers and broadcasters, or by TV viewers themselves.

Cultural Diversity: The variety and richness of different cultures and identities that exist in a society or in the world.

Impact: This refers to the significant effect that one event, action or situation can have on another. It represents the resulting change or outcome, which may be either positive or negative. Influence is the gradual capacity to shape opinions or behaviours over time, whereas impact is the immediate, measurable effect resulting from a specific action or event.

Influence: This term refers to the effect that one factor (variable) has on another. It is how changes in one aspect can cause shifts in another, based on research data.

Media Literacy: The ability to access, analyse, evaluate, and create media content in various forms and contexts.

Religiosity: The degree of commitment, involvement, or adherence to a certain religion or religious beliefs.

Social TV: The use of social media and digital technologies to enhance and extend the TV viewing experience.

Technology Awareness: This simply refers to an individual user's state of being informed about new gadgets, digital tools and innovations (like social TV), understanding their purpose and how they work, and feeling confident in using them effectively in daily life.

TV Consumption: The process and behaviour of watching and engaging with TV content.

UTAUT2: A theoretical model that explains and predicts the factors that influence the acceptance and use of technology by consumers.

Variable: A variable is any characteristic or factor in research that can change or be measured. It can be manipulated by researchers or observed to study its effect (influence) on other factors.

Chapter 4
Social Media Usage and Writing Skills Competence Among College Students

Alfred Okoth Akwala
Technical University of Kenya, Kenya

Joel Ngetich
University of Kabianga, Kenya

ABSTRACT

The study was driven by the need to investigate the use of social media and its influence on how college students engage with each other; and their writing activities and projects. Writing is a key component of communication competence whether in interpersonal, workplace or audience engagement in media productions. Communication engagements are shaped by the medium and this is evident in a society centered around digital mediums that are continually changing and advancing in their features. The exponential growth of social media and the high rate of mobile phone penetration has led to platforms such as Facebook, Twitter, TikTok and WhatsApp being a key part in youths' daily communication; yet these social media platforms require users to express themselves using limited words complemented by memes and acronyms. This paper adopted a

DOI: 10.4018/978-1-6684-9356-4.ch004

qualitative. Findings indicate that social media shapes the way students write as illustrated by the inconsistencies in their formal writing which is influenced by the informal writing, slang, emojis and acronyms used on social media.

INTRODUCTION

Communication, as a social process, is constituted in an interwoven and ever-evolving spatial and temporal framework. The arrival of new media and the continuously emerging technologies have drastically affected the nature of discussions on audience participation and media. The framework of audience engagement has then shifted towards audience participation as opposed to audience reception. The introduction of the internet and digital media technologies and human interactions have changed significantly in the last couple of decades. Digital media connects people in ways never before possible, enabling users to maintain friendships across time and distance. It enables those who are socially isolated or somehow set apart from their immediate physical community to connect with like-minded or like-situated people. It also facilitates interaction across social, economic, cultural, political, religious and ideological boundaries, allowing for enhanced understanding.

The development of the Internet and web 2.0 has created a whole new world of communication and today, social media has become a great social phenomenon especially among young people. For instance, Web 2.0 facilitated the feedback loop in online communication amongst members of a society by providing interactive communication platforms for different users to post comments, share posts or like posts. Companies use blogs as an informal, unstructured communication tool that can engage current and future customers and members of their wider community. Blogs are often used to start and maintain relationships, to update an audience, discuss new functionalities, and to inform about company decisions.

Social media is an interactive platform where users are able to generate contents based on mobile and web technologies for multiple sharing activities. Valerie and Campbell (2020) aver that technology is an integral part of daily life for many young people as much of their communication is mediated by digital and social media. They add that youths connect with each other through social networking platforms that allow for short, snappy messages, such as Snapchat with images that disappear after a few seconds.

Technology has been used to promote education through adopting different forms and modes of writing and participation in the digital age. Wil, Yunus, & Suliman (2019) state that the integration of social media is the latest medium that is used as an innovative way to delegate education purposes. Through adoption of internet languages, often derived and used within specific digital groups, social media users can employ communication technologies to attain various goals such as connecting with old classmates, meeting new friends, and sharing career updates. To achieve these goals, users often attempt to engage each other by posting updates and sharing links, and in doing so draw slang terms and codes that may at times only be understood between and among themselves. This kind of informal writing has always been perceived to have an influence on the formal writing style in terms of a language's syntax and semantics. This article seeks data collecting from university college students to generate discussions and draw conclusions on how informal social media language influences students' formal writing.

BACKGROUND TO THE STUDY

Social media is built on the idea of how people know and interact with each other. It gives people the power to share, making the world more open and connected with each other. Its growth over the past few years has changed the ways in which the internet is experienced by most end users. Faizi, Afia & Chiheb (2013) say that in the past, the web was a one-way information transfer tool as content was predominantly produced by experts, who published factual information and had the skills to create web pages. Users could only view websites but could not correct, change or add any content. Currently, the web has become an interactive and participatory "read-write" universal channel where users are allowed to be both producers and consumers of digital content in real time right through a web browser.

Kanana, Orwenjo and Gathigia (2022) explain that whilst the digital space is growing tremendously, human language is evolving too, to meet demands of the digital era. Most people use emojis, GIFs, logograms, pictograms, acronyms, abbreviations, neologisms, initialisms, emoticons, omitted letters, shortenings, and nonstandard language to communicate. They describe examples such as smileys and other symbols which express feelings of happiness, sadness, anger, or disappointment that have gained significance among the youth. Users have woven these networks into their daily routines, whether it is gaming, streaming movies, creating media, making friends, dating, attending social events and sustaining existing relationships. Digital media connects people in ways never before possible. It enriches people's relationships and their ability to stay in touch, and the ability of friends and family to help their loved ones in need through fundraising conducted on WhatsApp. It facilitates interaction across social, economic, cultural, political, religious and ideological boundaries, allowing for enhanced understanding. Faizi, Afia & Chiheb (2013) posit that social media encompasses a wide range of tools that integrate technology, social interaction and content the most popular social media types. These include social networks, blogs, micro-blogs, wikis, bookmarking, media sharing and RSS.

The rapid proliferation of social media has enabled availability of various internet applications which can be easily accessed allowing users to converse and interact with each other, to create, edit and share new forms of textual, visual, and audio content. It has a vital influence on the users lives and is widely used in politics, economics, and education. Stieglitz & Dang-Xuan (2012) say that social media provide a connection between social networks, personal information channels, and the mass media. They further argue that social media data in the form of user-generated content on blogs, microblogs, SNS, discussion forums, and multimedia sharing websites (e.g., YouTube) present many new opportunities and challenges to both producers and consumers of information. According to Kietzmann, (2011) social media is a platform that employs mobile and web-based technologies to create highly interactive spaces via which individuals and communities share, core-create, discuss, and modify user generated content. It has become an important space where users have opportunities to connect and identify with each other as well as manage their daily lives through selected social interactive activities.

The existing social media environment attracts different types of users and there are several reasons why people want to be on social media such as keeping up to date with latest events, redefining identities, maintaining friendships, gaming, dating, shopping (Facebook & Instagram), and creating new content for their platforms. However, while Internet connectivity is now far reaching and available in many schools and libraries, the technology is seldom employed in the rich and meaningful ways that youth employ technology in their literacy lives to dialogue with others with and through a variety of multimodal text forms (Warner, 2016).

Faizi, Afia & Chiheb (2013) describe a social network as an online community that brings people with common interests, opinions, activities and experiences together by sharing their news, photos, videos, and events. They further add that a majority of internet users are students who use social networks mostly to keep in touch with friends, colleagues, peers and family members.

Social media has also become a tool of instruction in the classroom, and one is likely to find teachers and college professors designing projects around YouTube, Twitter, Instagram and other social media applications to share students work, connect students with communities, nonprofits and foster collaborative projects. Wil, Yunus, & Suliman (2019) acknowledge that most pupils in the school are identified as the competent users of social media especially Facebook, Instagram and WhatsApp. However, based on the survey that was conducted shows that they engaged with social media items mostly for fun rather than for academic purposes. They add that therefore the use of Facebook and other social media items apart from educational usage are identified as a factor that helps to damage their academic performance.

Most social media users identify with specific groups of their choices on social media spaces and collaboratively engage in creating and modifying internet languages. The internet languages then create new forms of writing which users, especially younger generations use for communication among their social media groups (Hashin, Yunus, Ibrahim, Jeri, Sukr,

Ilahi, and Hasim, 2018). It is therefore becoming increasingly important to recognize and examine digital literacy practices of youth, as internet access expands to include more young people globally (Warner, 2016).

Social media as a communication platform offers users unique profiles, public connections, and the ability to create and consume user generated content. Hashin, Yunus, Ibrahim, Jeri, Sukr, Ilahi, and Hasim (2018) maintain that these relationships are sustained by a commonality in language and users draw from internet jargon, also referred to as new internet languages to develop an identity, build and maintain relationships and share regular updates with friends and followers. On the other hand, Aggarwal (2011) states that social networking sites are more about the tools used to make that content available to others and to allow users to connect, to engage with it and to build online communities. Social networking sites can also be defined as internet based social spaces designed to facilitate communication, interaction, collaboration, and content sharing across a network of contacts. Aggarwal (2011) further states that sites which are used for sharing online media content such as Flickr, Youtube, and Instagram can also be considered as indirect forms of social networks because they allow an extensive level of user interaction. Therefore, social media provides users means to express their personalities and views in a public forum. They provide many ways for users to express their personalities and views including applications and the posting of photographs and other content.

Social media has also enhanced information sharing across the world, giving people much greater access to facts, figures, statistics, and similar, allowing that information to circulate much faster. This not only enables people to respond in real time as events unfold but also helps to expose political corruption and unfair business practices. One group that is increasingly exploiting social media platforms is youth, most of them being the category of college students. The urge for college students to participate actively on social media platforms is higher compared to other groups. There are multiple reasons that attract different groups to register on social media platforms. These groups could best be categorized according to age, gender, occupation, level of education, social class, economic status as well as cultural reasons.

Social media also provides opportunities for users to engage in self-presentation of identity and communicate using "informal talk" with family, friends, and colleagues. Unlike traditional social network research examining face-to-face relationships, the focus of social media research remains on individual-to-individual communication rather than on relationships or connections. Online social interactions fundamentally differ from face-to-face interactions in offline social networks. Online social networking typically offers opportunities for expression that are less restrictive this is because users may self-disclose at levels quite different than in face-to-face conversations. While online social networks differ from face-to-face social networks in important ways, they nonetheless function as a viable channel for interaction. This dependence is due to the tools that social networking sites provide, allowing users to create and maintain social relations easily without undue constraints or fears created when it comes to direct face-to-face interactions in offline social settings.

Social media sites such as Twitter, Facebook, Instagram, Linkedin, Whatsapp and blogs commonly used by young people have connected users from different parts of the world and shifted the physical traditional spaces of communicative interactions to virtual spaces which have now been adopted by most users, especially the young generation. Social media among this generation has benefits spanning social engagement to include learning new skills. While social media sites are not academic sources, video sharing websites like YouTube assist students to write better and improve their global and cultural knowledge through self-learning. Other positive impacts are overcoming anxiety, generating new ideas for their essay writing and correcting their pronunciation (Hashin, Yunus, Ibrahim, Jeri, Sukr, Ilahi, & Hasim, 2018).

While social media has brought about numerous benefits, it also poses challenges for language and communication. This is in regard to the nature of the brevity and informality of communication which can at sometimes lead to misunderstandings, misinterpretations, and the spread of misinformation. It has transformed linguistic patterns, facilitated the emergence of new communication practices, and reshaped the way we interact with others. While critics argue that internet languages have a great impact on the writing skills of students especially in formal set ups, there has been no adequate research carried out in Kenya to establish how informal internet language use affects students writing skills competence. This article therefore seeks to address the following questions: a) What do students in higher institutions of learning use social media platforms for? b) What language do students in institutions of higher learning use on social media platforms? c) In what ways does social media use affect formal use of language among students in institutions of higher learning and d) What is the effect of social media use among students in institutions of higher learning?

USE OF SOCIAL MEDIA PLATFORMS BY STUDENTS IN INSTITUTIONS OF HIGHER LEARNING

Social media has become an everyday reality that changes many areas of life and the perception of self, and it is no longer possible to overlook or not accept (as is often the case in schools) social media in the structures of learning, because it has such a strong influence on the everyday life of pupils, and it automatically deforms the structures of interactions and thinking. In the education field, this media enhances learning opportunities for students to have wider scope of interactions with their teachers. A wide range of social media tools have been found to present many educational advantages for both students and instructors, hence providing opportunities for further learning.

Gudaitytė (2023) argues that Social media has become an everyday reality that changes many areas of life and the perception of self, and it is no longer possible to overlook or not accept (as is often the case in schools) social media in the structures of learning, because it has such a strong influence on the everyday life of pupils, and it automatically deforms the structures of interactions and thinking.

Social media has made it possible for one to create and easily transmit content in the form of pictures, videos, audios and words. It has influenced all spheres of life, both social, political, economic and educational. Faizi, Afia & Chiheb (2013) say that Facebook, MySpace and Twitter, for instance, can serve as backchannels for communication among students and between teachers and students within or between classes. Instructors can answer students' questions via a Facebook page or Twitter feed, post homework assignments and lesson plans, extend in-class discussions, send messages and updates, schedule or announce forthcoming events, and inform learners about special lectures, panel discussions or guest speakers. This teacher-student interaction will certainly contribute in understanding various learning-related difficulties and solving them in less time.

Successful learning depends on effective communication between teachers and their students. For this reason, teachers have to be in constant touch with their students in order to have an understanding of the students' problems, fear, or confusion which may inhibit successful learning. Excellent connectivity between the teachers and the students ensures students' optimal learning.

Faizi, Afia & Chiheb (2013) posit that a lot of educational institutions complain nowadays of their students' erratic behavior and poor scholastic performance. They add that if there is not enough communication between teacher and students, the feedback process remains thin or vague and optimal learning is not achieved. They further add that internet users, a majority of which being students, use social networks mostly to keep in touch with friends, colleagues, peers and family, it would be important to explore the possible benefits that such networking tools can offer to a student's learning. Wil, Yunus, & Suliman (2019) argue that social media can bring positive effects to promote enjoyment and facilitates the writing process where most of them use social media to learn English language and believed that social media is the best platform to develop better writing skills in English. Likewise, Faizi, Afia & Chiheb (2013) opine that using social media tools such as Google Apps enable students to have access to valuable learning resources regardless of time and place thus enhancing the learning experience.

Saba & Roy (2024) maintain that writing is a challenging task for every tertiary student because it expresses our thoughts, and what we are going to write is to discipline our language from academic to professional. They further add that students in tertiary education need to have strong writing abilities. Students often cite traditionally conducted classes as being boring due to lack of interaction with their teachers. Social media tools could be used to design learning activities that are both social and interactive. Besides promoting student-student interactions, social media can also be used to enhance and increase the number of interactions students may have with their teachers by overcoming the barriers of time and location. Consequently, students would have enough opportunities to ask questions, make comments, as well as get feedback. As such, they will have more chances to develop supportive relationships with their instructors.

College life focuses on academic and social community; joining interest-oriented communities such as on Facebook can provide a lonely student with a strong sense of belonging. Maintenance of relationships online has become common among college students as they use these avenues to create subgroups every day where individuals can find acceptance and camaraderie among like-minded users. Faizi, Afia & Chiheb (2013) argue that Facebook, MySpace and Twitter can serve as backchannels for communication among students and between teachers and students within or between classes. Instructors can answer students' questions via a Facebook page or Twitter feed, post homework assignments and lesson plans, extend in-class discussions, send messages and updates, schedule or announce forthcoming events, and inform learners about special lectures, panel discussions or guest speakers. This teacher-student interaction according to them will certainly contribute in understanding various learning related difficulties and solving them in less time.

Social media can improve communication not only between students and teachers but also among students themselves. Such as students talking to each other about upcoming assignments or tests or getting details from their classmates about what materials they could not understand in class. Sometimes students who have trouble with certain topics go to their classmates on a social networking website to get assistance and catch up online. However, some students seek entertainment needs like playing games downloaded or signed up for online and listening to music, watching movies by exploring YouTube, Netflix and other platforms. Wil, Yunus, & Suliman (2019) add that most pupils in the school are identified as the competent users of social media, especially Facebook, Instagram and WhatsApp. However, based on the survey that was conducted results show that they engaged with social media items mostly for fun rather than for academic purposes. Consequently, they conclude that the use of social media as an aim to assist pupils in writing skill has shown both positive and negative aspects. Therefore, teachers should conduct their classes through online platforms by having group work and discussions so that they can observe, guide and assist pupils in formal writing. These kinds of platforms offer them opportunities to learn and develop their talents as well. Others use social networking sites for socializing through chatting, playing games (competing), keeping in touch with family, maintaining communication with existing friends, sharing interesting things and updates, following a group, and expressing opinions on ongoing public debates.

While online social networks differ from face-to-face social networks in important ways, they nonetheless function as a viable channel of interaction for college students. Scholars maintain that never before have so many people gathered to communicate through a single medium.

Social media not only helps students to acquire knowledge but also establishes enduring relationships with real people. Connecting with fellow classmates through Facebook, Twitter and various social sites can help a student overcome the kind of isolation that otherwise might lead them to hate school. For instance, a Twitter account can provide a shy student with information about campus events that facilitate face-to-face encounters with other students. Such personal interactions are vital to creating and sustaining a sense of belonging.

The strength of social media applications is that they offer an assortment of tools that learners can mix and match to best suit their individual learning styles and increase their academic success. Social networking sites focus heavily on building online communities with common interests or activities. Social networking sites also can help students develop leadership skills, from low-level planning and organizing to activities that promote social change and democratic engagement. Social media tools and networking sites encourage students to engage with each other and to express and share their creativity. Such personal interactions are vital to creating and sustaining a sense of belonging. Kanana, Orwenjo and Gathigia (2022) argue that features characterizing African urban youth languages include: extreme multilingualism and lexical innovation; originating from argots; marking modernity and urban-ness; borrowing from or based on colonial languages but manipulating syntax and morphology; they have an associated extralinguistic style involving wider communicative strategies such as clothing, gestures, body language, ways of walking that serve to communicate modern street wise identities and urban sophistication.

As much as we have positive effects of social media there are a host of negative effects. One common negative effect posed by critics of social media is that it exposes students to a wide range of texts that may not be formally written according to grammatical rules of the language of use. This is evidenced by emerging grammar problems where students tend to use internet slang, abbreviations to keep writing short hence influencing their formal writing (Hashin, Yunus, Ibrahim, Jeri, Sukr, Ilahi, and Hasim, 2018). As such, Saba & Roy (2024) argue that traditional writing methods provide a strong foundation in grammar, syntax, and fundamental writing skills. Writing with pen and paper can enhance memory retention and understanding. However, they add that relying too heavily on technology might hinder the development of basic writing skills if not used strategically.

Students interact with an immense quantity of new phrases, words and acronyms which may not exist in the language of communication thereby interfering with their formal vocabulary. While the use of acronyms has been synonymous with our daily life especially when writing messages on paper or electronic devices, social media acronyms are constantly being invented and modified. Wil, Yunus, & Suliman (2019) note that writing rules on social media platforms are informal, thus it is undemanding for users who are not consummate in the English language to deliver their thoughts and ideas. Social media users are likely to use short forms in their communication such as Direct Message (DM), Ask Me Anything (AMA), LOL (Laughing out Loud), OMG (Oh My God), FYI (For Your Information), Be Right Back (BRB), In my Opinion (IMO), Throwback Thursday (TBT), Thank God It's Friday (THIF), Fear of Missing Out (FOMO), and others that keep being developed regularly on social media (Hashin, Yunus, Ibrahim, Jeri, Sukr, Ilahi, and Hasim, 2018). Investigating how individuals learn social media outside of formal learning environments can assist understanding effective learning practices and strategies on social media and how they translate into formal writing in the classroom.

TECHNOLOGY ACCESS, LANGUAGE AND PARTICIPATION

Advancements in technologies, development of the digital world and a myriad of digital cultures have provided opportunities that are highly embraced by youth, particularly university students. Online media and participation contain a wide variety of articulations of key concepts of access, interaction, and participation. Ordinary users are seen to be enabled (or empowered) to avoid the mediating role of the "old" media organizations and publish their material (almost) directly on the web, on self-managed websites and blogs, or on social media. The widespread use of the Internet, new integrative technologies and the new media have created a globalized world of virtual spaces that have changed the way people do interact and engage in their daily activities, business and even religion. Faizi, Afia & Chiheb (2013) say that students use social media as a tool of collaborative learning as they serve as platforms for learners to gather and share information and resources from both internal and external collaboration networks. Online learners, who could be students, educators, or any other individual in the community, can thus generate their own learning content and take advantage of collective knowledge. In this sense, social media are an effective means to create channels of collaboration between students and teachers and amongst students.

Meanwhile Kanana, Orwenjo & Gathigia (2022) are of the view that modern youth access digital spaces which are not limited to geographical space – neither local nor rural – to keep in touch with global trends and culture, manifested in the language. They also add that trending global fashions, movies, music, and stories are easily adopted and localized through language, images, and memes, albeit with a local, contextualized flavour. The impact of this new media on society varies in both positive and negative outlooks. Internet makes everyone a publisher and everyone a librarian, in that anyone can both produce and retrieve an unprecedented amount of information. Faizi, Afia & Chiheb (2013) say that a student who hardly ever participates in class may get actively engaged in co-constructing his learning experience with his teachers, collaborating with his fellow colleagues, and may feel more comfortable expressing himself and sharing his resources and ideas on Facebook, Twitter, or YouTube.

Social media tools are also effective ways to increase students' engagement. As the social features of social media resources have attracted the attention of millions of people around the globe, these same features are also learning opportunities provided by their academic institutions. The use of social media in educational institutions helps students, teachers and parents to get more useful information, to connect with educational systems that make education convenient. This ensures that institutions and students access multiple opportunities to improve learning methods.

LANGUAGE

Language is the connecting tool for communication between users of social media and hence the various categories and varieties of language used determine to a larger extent the social group of the participants. Language may be used formally or informally as it is a dynamic communication tool that evolves to accommodate new words, signals and nonverbal expressions wherever users evolve, and the same case applies to language use on social media. A systematic review of vocabulary research done by Hairrell, Rupley, & Simmons (2011) stated that targeted vocabulary learning leads to an increase of word knowledge. The study found that repeated exposure, contextual analysis, and semantic strategies are three common methods in building up vocabulary. According to Nagy & Heibert (2011) vocabulary development is also influenced by wide exposure through reading. Frequency of time spent in wide reading, language proficiency, and text complexity are factors that affect the relationship between wide reading and vocabulary development (Ford-Connors and Paratore, 2015).

Due to technology, words like blogging, tweeting, facetiming, snapchatting and trolling have been rampantly used by social media users including college students (Hashin, Yunus, Ibrahim, Jeri, Sukr, Ilahi, and Hasim, 2018). These informal words might accidentally find themselves in formal communication contexts due to frequency of use. Mworio (2015) opines that English language users who are not versed with the modified words used on social media platforms may unknowingly apply the new terms they come across on social media in formal writing assuming the words are standard English words. Academic writing is increasingly facing concerns because students are increasingly writing formal papers in a more casual tone (Hashin, Yunus, Ibrahim, Jeri, Sukr, Ilahi, and Hasim, 2018).

Social media helps in connecting learners with individuals across the globe, but as stated by Abbas, Aman, Nurunnabi, & Bano (2019) the more time learners allocate on social media, the less time they spend interacting in person with real people, and it subsequently impacts their communication skills. Writing is one of the major skills affected by over exposure to social media. Although Facebook, Instagram and YouTube have been recognized as part of learning mediums, these platforms have negative effects. As an example, Facebook is an online space where users can make use of writing and express themselves, while being exposed to many written posts, however according to Derakshan & Hasanabbasi (2015), many of the posts are informal and are not considered as forms of academic writing that learners can learn from and apply in language learning settings. The emerging concern now is that writing associated with texting and social media is now infiltrating academic writing and creating issues with spelling, grammar, sentence/paragraph structure and formality (Hashin, Yunus, Ibrahim, Jeri, Sukr, Ilahi, and Hasim, 2018).

THEORETICAL FRAMEWORK

An appropriate theory in research acts as a guide to a researcher and the lens through which the research gains insight into the phenomenon under study. Littlejohn and Karen (2008) say that a theory introduces one to new ideas and helps them see things in new ways (p. 28). Therefore, theories guide one in trying to discover what they are investigating as it opens the eyes to new things. They guide an investigator to move from curiosity to observation and lastly draw a valid conclusion about a given circumstance. Theories hence assist a researcher to adapt the findings to solve real problems in society. There are several theories that can be used to investigate and bring to focus social media and writing skills issues. These include Social Identity Theory (SIT), Uses and Gratification Theory (UGT),The Social Capital Theory, The Network Effects Theory, Spiral of Silence Theory, Social Exchange Theory, Diffusion of Innovations Theory, The Cultivation Theory, Social Information Processing Theory (SIPT), Mashall McLuhan technological determinism theory among others.

This article, however, draws from two theories to frame its arguments, methodology and discussions. The Social Information Processing Theory (SIPT) proposed by Joseph Walther (1992) and Mashall McLuhan technological determinism (1964) which argues that the way we communicate, through different media, alters our modes of thinking and shapes our social structures.

The social information processing theory explains how individuals use computer-mediated communication to develop interpersonal impressions and focuses on how communicators adapt in the absence of nonverbal cues when restricted to textual symbols. Social media users are attracted by the new words and phrases that are created and used on social media. On the other hand, Marshall McLuhan's theory of technological determinism (1964) suggests that technology shapes human perception and cognition and hence evolution of writing technologies such as social media have profound impacts on writing styles, cognitive structures, and the overall communicative environment. McLuhan argues that technology, particularly communication technologies, significantly shapes human thought, culture, and society. He continues to explain that when new systems of technology are developed, the culture or society is immediately changed to reflect the senses needed to use the new technology. The theory predicts that with every new system of media technology, society will change and adapt to that technology.

Gudaitytė (2023) quoting McLuhan (1964) states that social media is a game-changing technology because of the impact it has already had and is still having on changing everyday life, the ability to change the pace of communication and the transmission of information (things are speeding up), and the habits of mind. We expect to receive information much more quickly, the emotions of live communication are synthesized in 'emojis', the creation of a sense of proximity at a distance, the promotion of anonymity and individuality through a presumed sense of community, which manifests itself in the possibility of secretly 'meddling' in the affairs of others.

Gudaitytė (2023) says that the speed of social media also poses certain challenges that require teachers to acquire new competences and to teach their students not only the content, but also how to operate in the new media and what messages the media itself conveys. The available speed comes at a cost of the accuracy and reliability of the information provided online, which also poses a challenge for the education system. He further maintains that social media used for educational purposes are not inherently bad, and they are also in line with the positive aspects of the Global Village, but the importance of discussing ethical norms and promoting the aforementioned critical thinking of students by drawing attention to the influence of the media itself on their thought processes and interaction with their environment increases. He thus concludes that looking at social media in education through McLuhanian lens, the school is at a crossroad – how and whether to incorporate social media, knowing the dangers McLuhan predicted, and how to teach the young person to critically evaluate the social media that have become inevitable, and the specific messages they carry, as well as the ability to shape the message independently of the content.

METHODOLOGY

A qualitative approach was adopted in this study. Lindlof &Taylor (2002, p. 66-7) argue that qualitative researchers are disciplined; must learn when to watch, when to listen, when to go with the action, when to reflect and when to intervene tactically. Jwan (2010) defines it as a flexible approach that seeks to generate and analyze holistic data on an issue of interest using sufficiently rigorous, trustworthy and ethical methods and techniques. In this study, the qualitative approach enabled the researcher to gain deeper insight into social media communication.

The collaborative and comparative approach enabled us to sample diverse students' responses that were generated during the study. This article was a result of a joint project between two professors, one teaching a first-year core course in communication and writing skills and another one teaching advanced multimedia writing and film communication. It allowed us to collaborate on the diversity of writing and how routine use of social media slang, emojis and memes intersects in students writing research papers, creative writing, and multimedia writing.

The target audience were students selected from two institutions of higher learning, an institute of Mass Communication offering creative writing courses at various levels and a technical university offering general writing courses for first-year students in communication skills with a section on writing skills. The sample population was thirty (30) students from the selected two institutions of higher learning. Fifteen students were sampled from each institution and five from each institution requested to share scripts from selected WhatsApp conversations anonymously with personal identifying information and photographs redacted. The remaining ten from each institution were requested to undertake a written exercise, using pseudonyms to provide written data. The exercises were divided into two groups, an essay and a creative writing exercise.

The sampling procedures and data generation methods used were appropriate for this study as they allowed the researcher to investigate a specific set of issues through the more structured qualitative approach. The main objective of the study was to interrogate the effects of social media on the writing competency among students in institutions of higher learning.

The study was guided by the following specific objectives:

(a) To determine various ways in which students in institutions of higher learning use social media platforms for communication.
(b) To establish the kind of language used by students in institutions of higher learning on social media platforms.
(c) To examine the various ways by which use of social media affects students' writing skills.

DATA PRESENTATION AND DISCUSSIONS

The data was generated from two stages. The first stage involved social media conversations shared anonymously by the participants and the second stage included participants engaging in handwritten exercises, using pseudonyms. The data generated from the two stages of data generation was intended to identify any anomalies that may be experienced due to exposure to social media.

An analysis of the various ways in which students in institutions of higher learning use social media platforms from WhatsApp conversations shared anonymously revealed that students use social media to engage with each other on various topics such as family, relationships, finances, school-related academic and social events. WhatsApp provides many ways for users to express their views including applications and posting photographs. The lack of bodily gestures and other nonverbal cues, like tone and inflection on social networking sites has been replaced by emojis and punctuation, still not an adequate replacement for face-to-face communication. Students who spend a great deal of time on social networking are less able to effectively communicate in person. It also emerged that students, who attempt to multi-task, checking social media sites while studying, showed reduced academic performance as well as excellent writing.

Many students in institutions of higher learning rely on the accessibility of information on social media and the web to provide answers. The amount of time students spend on social media was an emerging concern among them, and several of them indicated that their family members were worried that the more time they spend on social sites; the less time they spend socializing in person with family and friends. The prime concern of social media use among college students then is addiction. A concern that students get addicted to social media to a worrying level of them forgetting to communicate with offline networks or working on their class assignments. This addiction is therefore negatively affecting college-level valued activities such as concentrating on studies, taking active part in sports, offline campus events and communication with family and friends outside of social media.

An analysis of the kind of language used by college students on social media showed that WhatsApp serves as an online space in which social relationships form and are sustained by user's ability to read, write and share short informal messages and media with family and friends. However, an overly undue reliance on this kind of writing has an effect in cases where students are expected to be formal in their writing. It emerged that the ethics of writing and communication have become blurred due to the degree at which private information is self-disclosed and shared online due to assumed high levels of anonymity on the internet. This then manifests in academic writing where students forget the need to filter the information they share in their written assignments and academic essays.

From the findings, there are various wide-ranging impacts of social media on students' relationships, social skills, communication, mental health and more specifically for this article, students writing skills. Students indicated that they were having a harder time getting to communicate face to face with people, and were losing their social skills, due to the fact that they were spending more time behind a screen. Students also pointed out that the amount of time spent on social media was affecting their writing, as they were forced to spend more time trying to remember the formal spellings of words. This was due to shortened forms of writing they used on social media and so the same kind of language used on WhatsApp was reflected in their writing assignments. The analysis indicated that the effect of using WhatsApp by students has a significant effect on their writing. The discussions below focus on the impact of social media on students' writing skills.

SPELLING ERRORS DURING WRITING

The popularity of social media, and the speed at which information is published, has created a lax attitude towards proper spelling and grammar. This was evident during independent handwritten exercises, where students faced several challenges especially in connection to spelling and general communication. This then showed an increase in students' overreliance on the computer's spell check feature. From the conversations shared from WhatsApp it was also evident that writing on social media was affecting students' spelling abilities.

> *I do **bealive** when it comes to love it is not always up to us but it is up to the fate that makes us meet(S1/Earnest)*
>
> *There are many forms of entertainment, for example partying, **chilling with friends and vibing with oneself**(S2/Nikol)*
>
> *Education is the **knoledge** gained from institutions such as schools. Education helps and equips one with skills required in the field(S3/Annah)*

Misspelled words Correct form

1. Sorry for the inconviniens Sorry for the inconvenience
2. She got into an arguement She got into an argument
3. That is not very comon That is not very common
4. Do not embarras me Do not embarrass me
5. I will adress the issue later I will address the issue later

CODE- MIXING

An analysis of WhatsApp conversations revealed that a lot of students end up mixing Kiswahili and English in their communication. This is a common form of communication in Kenya, particularly the use of sheng among young people. From the data, it was evident that sheng was predominantly used and code-switching where participants mixed English and Swahili words when communicating. This was evident in the writing assignments where there was an element of direct translation from Swahili to English. Examples of code-switching can be seen below as generated from the WhatsApp conversations.

> **Heee, lemme** *tell you. So jana after getting my nails done. I was to go to the movies with* **my date**. *Alafu he called me and said he's getting his hair done...*
>
> *Anyway first forward, tulitoka hapo 5pm. Nikona njaa na* ***tumemiss*** *movies (S4/Nick)*

Code mixed language Pure language

1. Tomorrow tunameet where? Where are we ***meeting*** tomorrow
2. Si we can do it kesho We can do it ***tomorrow***
3. Amezima phone ***He/she has switched off*** the phone
4. Edit hiyo document nimekusend Edit ***that*** document ***I have send you***

USE OF INFORMAL WRITING

There is an increase in use of informal writing often found on social media by students in their college essays. The usage of internet slang is very common in social media as the language is easier, instant and short which saves time when communicating online (Hashin, Yunus, Ibrahim, Jeri, Sukr, Ilahi, and Hasim, 2018). The challenge is that the writing skills taught in class are faced by the increase in the lack of ability to retain information driven by social media. Therefore, the willingness to spend more time researching and looking up good information has reduced, in this case mastery of formal writing in that students are used to the ease of accessibility to information on social media, often written informally. Below are examples of informal writing used by college students.

Informal Writing Formal writing

1. Let me call my bro Let me call my ***brother***
2. We can link up next week We can ***meet*** next week
3. I wanna go to class right now I ***want to*** go to class now
4. Could you scoot for me a little ***Move over*** a little
5. Hit me up anytime mahn ***Call*** me anytime ***man***

USE OF SHORT FORMS

The more time students spend on social sites; the less time they spend socializing in person. This then affects their ability to express themselves, which then gets reflected in their writing skills, leading to an overreliance on use of short forms for communication. Social media users tend to leave out letters from a word to make it shorter to save time and limited characters on certain social media platforms (Hashin, Yunus, Ibrahim, Jeri, Sukr, Ilahi, and Hasim, 2018). …. The findings in this study revealed that social media played a big role in elevating informal language among multilingual users and subsequently influencing the use of formal language in both speech and writing.

Short form Equivalent long form

1. Idk, you find out *I don't know*
2. Wdym you are broke *What do you mean*
3. Hyd today *How are you doing*
4. We can go rn *Right now*
5. Wyd later tonight *What are you doing*
6. Hmu in the course of the week *Hit me up*
7. Aight I will get back to you *Alright*

AFFIXATION AND COMPOUNDING

Social media users make use of the social media terminologies in a modified way to communicate meaning that is understood within a particular group during interactions. These terminologies are affixed in such a way that they communicate a particular meaning to the users. For example, the following words were found in the students' writings. They include ***unfriend***, ***poking***, ***tagged***, ***tweet***, ***unfollow*** and ***unlike***.

Conversation one, two and three above further illustrates the influence of social media on students' writing. The above findings are also reinforced by Zimmer (2014) who asserts that young adult users of social media platforms make use of the informal terminologies and modify them to communicate meaning that is understood within a particular group during interactions. Another influence has been in following grammatical rules in word and sentence construction. In morphology, the method for creating new words from morphemes in English is using base words, prefixes and suffixes as shown in the examples below:

pay (base word) + pre (Prefix) = prepay (new word);

Enjoy (base word) + able (suffix) = enjoyable (new word).

However, language use on social media does not follow the productive rules of grammar, research findings indicate that social media users form words and sentences without following the productive rules of grammar either due to ignorance or neglect. Faizi, Afia & Chiheb (2013) say that given that a large number of internet users are teachers and students, social media seem to have greatly influenced the way we teach and learn.

In investigating how social media influences students' writing skills, there is a need to know the impact of social media on students' education especially writing skills and the impact it has on the next generation so that we make appropriate plans in both cases for appropriate usage of social media.

CONCLUSION

Writing well is an essential skill of academic success, and it is therefore imperative to appreciate the impact that writing has on students' careers and students' progression in the career path. The writing skills acquired by a student during his/her education can impact positively or negatively depending on how they share and obtain input through the written form.

Usage of social media differs from person to person as it has good as well as negative impact on users and society. Critics of social media argue that social media has become a major distraction to students, causing the overall performance of students to decline, especially the ones who tend to constantly check their Facebook, Twitter and TikTok while attending lectures or studying. While social media applications allow students to converse and interact with each other; and to create, edit and share new forms of textual, visual and audio content, it can also be argued that students' ability to concentrate on the task at hand is significantly reduced by the distractions that are brought about by YouTube, Facebook, TikTok or Twitter.

From the above discussions, it can be concluded that social media has various negative effects on students. The popularity of social media, and the speed at which information is published, has created a lax attitude towards proper spelling and grammar when using social media for interaction among students. Due to the reliance on social media, a student's ability to effectively write without relying on a computer's spell check feature is reduced. Social media has also become a key source of information and teachers and adults are no longer the only sources of knowledge. Salaudeen & Lawal (2019) proposes that further investigations on challenges posed by social media on writing culture should be considered among different categories of students, lecturers, educators, and student support services to aggregate their perspective and experiences on the relationship between addiction or heavy use of social media and students' writing abilities. As much as the advent of social media is appreciated, there is a need to relook at how it can be effectively used to help enhance communication in education settings and improve learning outcomes among students.

REFERENCES

Abbas, J., Aman, J., Nurunnabi, M., & Bano, S. (2019). The Impact of Social Media on Learning Behavior for Sustainable Education: Evidence of Students from Selected Universities in Pakistan. *Journal Sustainability*, *11*, 1–23.

Aggarwal, C. C. (2011). An Introduction to Social Network Data Analytics. In Aggarwal, C. C. (Ed.), *Social Network Data Analytics*. Springer.

Faizi, R., Afia, A. E., & Chiheb, R. (2013). Exploring the Potential Benefits of Using Social Media in Education. in *International Journal of Emerging Technologies in Learning (JET)*. https://www.researchgate.net/publication/272998596

Ford-Connors, E., & Paratore, J. R. (2015). Vocabulary instruction in fifth grade and beyond: Sources of word learning and productive contexts for development. *Review of Educational Research, 85*(1), 50–91. Retrieved May 5, 2019, from https://journals.sagepub.com/doi/10.3102/0034654314540943

Hairrel, A., Rupley, W., & Simmons, D. (2011). *The State of Vocabulary Research. Literacy Research and Instruction.* 50(4), p.253 271. Retrieved 10 June, 2019. https://www.tandfonline.com/doi/full/10.1080/19388071.2010.514036?scroll=top&nedAccess=true

Hashin, H., Yunus, M. M., Ibrahim, N. S. M., Jeri, I. E., Sukr, M. A. M., Ilahi, N. S. N., & Hasim, M. A. N. (2018). Social Media and Its Impact on Students' Writing Skills. *International Journal of Engineering & Technology.* 1(7): 102-106. Last accessed 24 sep 2022._https://www.sciencepubco.com/index.php/ijet/article/view/21624. doi:http://dx.doi.org/. v7i4.21.21624.DOI: 10.14419/ijet

Jwan, J. O. (2010). *Conducting Qualitative Research: Current Trends and Developments* Unpublished Paper presented at the Moi University 5th Campus Wide Research Workshop, Eldoret. 7th & 8th Oct, 2010. Print.

Kanana, F., Orwenjo & M, N.Gathigia, M.(2022).Escaping the Margins of Society: New Media and Youth Language Practices across the Rural Urban Divide in Kenya. *In Multilingual Margins 2022, 9(1): 84-101*

Lindlof, R. T., & Taylor, C. B. (2002). *Qualitative Research Methods*. Sage Publications.

Littlejohn, S. W., & Karen, A. F. (2008). *Theories of Human Communication* (9th ed.). Thomson Wadsworth.

Mworio, I. (2015). *Use of English Neologisms in Social Media:* University of Nairobi. M.A Thesis.

Nagy, W. E., & Hiebert, E. H. (2011). Toward a theory of word selection. In Kamil, M. L., Pearson, P. D., Moje, E. B., & Afflerbach, P. P. (Eds.), *Handbook of Reading Research* (Vol. 4, pp. 388–404). Longman.

Saba, Z., & Roy, S. (2024). Technological Impact on Tertiary Level Students' Writing Skills in the Post-COVID Era. *In International Journal of Research and Scientific Innovation (IJRSI).* ISSN No. 2321-2705 | DOI: |Volume XI Issue II February 2024DOI: 10.51244/IJRSI

Salaudeen, K. A., & Lawal, H. A. (2019). *Social Media Addiction and Formal Writing Skills Among Mass Communication Students*. Southwestern University.

Valerie, M., & Campbell, V. M. (2020). Digital Stories as Data. In Servaes, J. (Ed.), *Handbook of Communication for Development and Social Change*. Springer.

Warner, J. (2016). Adolescents' Dialogic Composing With Mobile Phones. *Journal of Literacy Research*, *48*(2), 164–191. DOI: 10.1177/1086296X16660655

Wil, C. S. C., Yunus, M. M., & Suliman, A. (2019). The Use of Social Media to Assist Writing Skills among Secondary Pupils. *International Journal of Academic Research in Progressive Education and Development*, *8*(3), 224–236.

Chapter 5
Social Network Sites Audience Engagement on Health Reports:
An Evaluation of Daily Nations' X Platform, @DNHealthyNation

Denish Ouko Otieno
https://orcid.org/0009-0002-3293-6973
Moi University, Kenya

Grace L. S. Mwanja
https://orcid.org/0000-0002-5145-4438
Moi University, Kenya

ABSTRACT

The news media play an important role in informing the public on scientific research findings in health matters. In Kenya there has been unprecedented rise in the use of Social Network Sites more so among the youth. According to a report by SIMElab Africa, 2019, X is specifically used more by those aged 26 to 35 years in addition to being used by journalists and media organizations to disseminate information. Condit (2004) posits that there has been a mismatch between the expectations of the scientists and those of the reporter on disseminating scientific research findings which

DOI: 10.4018/978-1-6684-9356-4.ch005

sometimes twists the messaging. The overall objective of the study was to establish framing of health research findings on X as a news media platform with focus given to The Daily Nation's Health and Science pull out '@DNHealthyNation'. The study found out that there is a mismatch between health reports on social network sites and original scientific health disseminations due to factors such as competition for readers' attention by journalists which leads to sensationalized report interpretations

INTRODUCTION

This study highlights how the public react and interact to scientific Health disseminations on Social Networking Sites (SNS) shared by news media. It further shades light on messaging of scientific research disseminations by journalists and the expectations of the researchers.

As highlighted by Aggarwal (2011), Social Networking Sites are more about the tools used to make content available to others and allow users of the tools to connect, engage with it and build communities. This definition is supported by that of McBride (2012) referring to SNS to be web built to allow people to express themselves and interact socially with each other.

While noting that the nature and nomenclature of connections may vary from site to site, Donah & Nicole (2007) define SNS as web-based services that allow individuals to (1) construct a public or semi-public profile within a bounded system, (2) articulate a list of other users with whom they share a connection, and (3) view and traverse their list of connections and those made by others within the system.

Aggarwal (2011) posits that types of Social Networks can vary based on the objectives of such sites. There are several types of Social Networks such as Profile-based and connection sites which are organized primarily around members' profile pages which consists of likes, interests, dislikes and pictures. Such sites include *Facebook, Bebo and MySpace*.

The second category of Social Network Sites are Multimedia Sharing Social Networks that allow users to upload and share video content easily. i.e *Youtube.* The third category of Social Network Sites are the Microblogging sites which allow users to exchange undersized content such as short sentences, video links, and images. i.e. X and Tumblr.com.

X is among the most popular social networks with over 500 million users around the world. It consists of a "microblogging' service that allows the sharing of short messages as well as including images, videos and hyperlinks to other sites (Goni et al. 2018). X is meant to produce, exchange and spread concise statements, making it one of the best platforms to share information in a very quick, efficient and collaborative manner. Goni et al. further states that X has become an important communication channel for stakeholders' groups in science as major academic publishers and key scientific journals have X accounts. In addition, scientific societies and research centers use X to promote their research disseminations. It is upon this backdrop that positions X as a key platform for the study. The HealthyNation X platform which is the Daily Nations Health and Science pull out @DNHealthyNation is such an example. A study by Portland Communications (2016) and Tweetminster, shows that most journalist use social media and specifically X to Tweet, Retweet, and share short bursts of information to their followers.

By the time of this study, the Nation Media Group through its X page #Nation Africa "*@NationAfrica*" had three million (3 million) followers. Its print division which comprises of Daily Nation Health and Science pull out X page "*@DNHealthyNation*" had three thousand seven hundred and fifty-six (3,756) followers. #Standard Group "@Standard Media" had one million, eight hundred followers (1.8 million). The data positions Nation Media Group as having the largest followers on X via #Nation Africa page. This therefore underscores why the Nation Media Group's, Daily Nation Health and Science Pull Out platform @DNHealthyNation was chosen for the study. In addition, the newspaper had a specific X page that deals with dissemination of science research on health matters. This niche helped in the study focus.

With advancement of technology that has led to the worldwide connectivity of computers, the internet gave birth to an online community. Media organizations are now making use of online platforms to reach out to the online community. Online communication technologies were introduced to the public in forms such as email and chat rooms in the early 1990's. This is documented by Peter & Valkenburgg (2009) who note that today online communication has evolved to be more interactive with multiple feedback given real-time as people network.

Social Network Sites can be embraced by news media organizations to disseminate health information. The sites are effective due to their timeliness and wide reach. According to Heldman, Schindler and Weaver (2013), social media networks have become an integral part of the public health conversation with number of users and voracity at with which the users consume information on this site graduating over time. Unlike other forms of mass media like print and broadcast media, SNS provide platforms for direct and active engagement, where one can obtain simultaneous information.

Scientific research addressing health issues play a critical role in providing relevant data and information that if shared with the public can be utilized to ensure healthy practice for the wellbeing of all. As described by the European Public Health Association (EUPHA), health reports communicate the state of health of a population 'to all who need to know' and to support the preparatory as well as evaluate phases of health policy making and health care planning. The Association observes that public health monitoring is useless without public health reporting.

Today, news media organizations are embracing SNS like X and Facebook as alternative platforms to their main news media. The Daily Nation Health and Science pull out X page # HealthyNation *"@DNHealthyNation"* is among the news media organizations in Kenya that have adopted SNS as alternative platforms to reach their audience.

Research done by Kamenchu Mwongelu (2012) on how social media have changed newsroom operations in Kenya highlights that social media plays a complimentary role to other media platforms. In the case of Nation Media Group (NMG) which Mwongelu research focused alongside the Standard Media Group, it was revealed that social media helped drive traffic to TV, newspaper and radio owned by the two media houses, and to generate debates around topical issues with audience and traditional sources. In this study, health reports were seen to be a traditional source for NMG news for their Daily Nation Health and Science pull out X page # HealthyNation '*@DNHealthyNation*'.

The study also inclined on X as SNS in the category of social media. As a SNS, social media offers many challenges when it comes to safety and consistency of news content. However, news media organizations are ignoring this and opt to use social media to reach a wider audience as this is core for their popularity.

Oketch et al (2016) notes that social media use in news broadcasting leads to privacy breaches, lack of policy frameworks, infrastructure security, legal and intellectual property issues, resource challenges and potential risks. These researchers however observe that despite the challenges in social media use, the opportunity to reach everybody, everywhere, at all times is too opportune for news broadcasters to ignore.

Recent research findings indicate that X is becoming an important source of information in Africa with 68 percent of respondents acknowledging to use X to monitor news compared to 22 percent that use it to search for employment and only 10 percent use it for other purposes (Portland 2022). This high rate of X use in Africa saw it to be the most relevant social media platform of focus this study.

In view of the above, this study focused on # HealthyNation X page for Daily Nations' Health and Science pull out '*@DNHealthyNation*' as a communication tool that provides a platform for users to connect and communicate, creating social communities within the page based on the health content shared.

Study Background

Public health reports are sources of important information that guide healthy practices and behavior and can also sensitize audiences to emerging health issues that require them to be cautious. The health reports can however be ineffective if original research information is distorted.

When health researchers come up with their findings, it is their intention that the findings reach out to their target audience to enable them to adopt healthy practices that lead to their wellness. However, studies indicate that there has been a mismatch between the expectations of the scientists and those of the reporter to disseminate scientific research findings, which twists the messaging (Condit, 2004).

Despite being a reliable means of reaching a wider audience in a timely way, among other advantages, SNS can lead to distortion of health reports as a result of freedom of participation which gives room to audience to give varied comments on the shared content, which might be different from the original report. The rapid adoption of the use of SNSs raises important questions about the social implications of their usage (NielsenWire, 2010). The varied information shared on SNS platforms may lead to negative perceptions which are obviously against the intention of health researchers. As sited by Poutintsev (2023), one of the challenges of social media platforms is misleading information since it gives room for publishers and spammers to share misleading information.

Understanding the penetration of news in social networking sites and in light of addressing the challenges that come with the use of these sites when sharing sensitive content this study evaluates X engagement on health reports on #HealthNation X page of the Daily Nation @DNHealthyNation.

Overall Study Objective

To establish framing of research findings on X as a news media platform and analyze how this affects understanding and behavior of audiences by accessing health reports through X as a Social Networking Site.

While Nation Media Group (NMG) has a X page #Nation Africa with 3 million followers, the study specifically focuses on the Daily Nations' Health and Science Pull Out X page #HealthyNation @DNHealthyNation.

Specific objectives of the study were to establish reaction and interaction of X users with health scientific disseminations news reports and to determine researchers' expectations on scientific disseminations on news media sites.

Study Questions

a) Are health reports on X framed the same way as they appear on original reports?

b) How does X audience react to health news reports disseminated on X platform @DNHealthyNation?
c) What are expectations of researchers on disseminated scientific health reports on X?

Theoretical Framework

Theories Linking Messaging and Groupings Based on Similar Perceptions

In this section, the focus is to highlight theories that tend to explain how messaging of information and communication media can affects perceptions and lead to groupings among audience with similar interests.

Symbolic Convergence Theory

Symbolic Convergence Theory (SCT) highlights how varying messages can affect meaning or perception of original information. This is with a focus on messages that are usually shared within a group. This theory is credited to Professor Ernest Bormann who between the year 1925 and 2008 worked in the department of speech communication at the university of Minnisota. His theory emphasizes the sharing of creative interpretations as a method of developing shared meaning. The theory perceives human to have a tendency of attributing meaning to signs and objects in order to make sense of them, just like highlighted in the attribution theory (Endres 1994).

The theory explains the appearance of a group consciousness, with its implied shared emotions, motives and meanings, not in terms of individual fantasies and scripts but rather in terms of socially shared narratives (Bormann 1985).

This theory has evolved over time with several researchers making contributions to it. Shields (2008) states that SCT is a general communication theory which explains the emergence of common symbolic consciousness which comprises of shared meaning, including emotions, values and motives for human participation within a small group.

In this study we look at how sharing of health reports on SNS affects original information. As explained earlier, X as a SNS gives room for audiences to comment on posts. This allows for different narratives to be given as comments on a post. The audiences come together to form groups based on their interests on posts shared.

For this study, we focus on Daily Nations' Health and Science Pull Out X page #HealthyNation *"@DNHealthyNation"* where one of the contents is news which is meant to inform the public on emerging issues and trends that may include health reports on emerging diseases, invention of new drugs, health campaigns, among others. Though a number of X users are attracted to Nation Media Group X page #Nation Africa '@Nation Africa", a specific niche of X users are attracted to the Daily Nations' Health and Science Pull Out page *"@DNHealthyNation"* to get information on scientific health breakthroughs. As they do this, they form groupings of audiences with similar interests based on their emotions, values and interests.

The Speech Act Theory

The Speech Act Theory (SAT) is a theory that highlights how words are important in creating messages in the communication process. The theory was invented by an oxford philosopher Austin (1975) while looking at 'How to do things with word', way back in 1960s. According to Austin, speech act is something expressed by an individual that not only presents information but performs an action as well. These actions can entail sharing different narratives and thus link it to the Symbolic Convergence Theory which acknowledges appearance of a group consciousness in terms of socially shared narratives. The theory is a subfield of pragmatic that studies how words are used to present information/messages and to carry out actions (Nordquist, 2024).

As indicted by Austin, speech act can be distinguished as performatives which are words that are used to undertake an action that is rather felicitous or infelicitous (e.g. he is a dog) and constatives words that are used to make just a statement which can be either true or false (e. g. you are a witch). X as a social media platform gives room for audiences to share performances since it offers its members freedom of participation. This the audience can do by giving either felicitous, infelicitous or constative words.

Austin further indicates that speech act can take place in two felicity conditions: Constitutive conditions- necessary to successfully perform a speech act. Regulative conditions- concerned how happily or how well it is performed.

The speech act theory was greatly improved by contributions from John Searle (1969) who added to the findings of Austin by highlighting that apart from people using language to assert things, they also do the things. Just like Austin, Searle believes that the meaning of a speech act cannot be accounted in the absence of the context and that context alone does not express propositions rather than sentences in a context or tokens, express propositions. Just like Austin, Searle indicates that there are many ways of describing same speech act (physical act, act of reference, elocutionary act, and illocutionary act). Searle divides the allocative acts into:

Representatives: these are words that contain a certain statement about the world e.g. describing, claiming, concluding… Example: `vaccines don't affect children like they affect adults.'
Directives: words that direct the hearer towards doing something, e.g. ordering, asking, requesting, advising… Example: 'please don't trust these health report.'
Commisives: Speaker commits himself to doing something e.g. promising, threating, offering… Example: `I will take the COVID-19 vaccination only after the President takes.'
Expressives: They express how the speaker feels about a certain situation, e.g. thanking, apologizing, welcoming… Example: 'Thank you for warning us about the vaccine.'
Declarations: The point of which is to bring something about in the world e.g. christening, marrying, firing Example: 'All those do not vaccinate against COVID-19 will be restricted from boarding Public Service Vehicles.'

All the conditions described in the SAT are evident in social media today including X. The theory was therefore relevant for this study which focused on identifying how information on #HealthyNation "*@DNHealthyNation*" X platforms affect perception on health reports that are shared on this site.

METHODOLOGICAL SCOPE

Introduction

It entails the approach, methods, techniques and procedures by which a researcher collects, analyzes and presents data based on objectives of the study (Serem, Boit, & Wanyama, 2013).

Qualitative research approach was used in the study. This was informed by our philosophical assumptions, data collection methods, analysis and interpretation. As argued by Creswell (2014), the selection of a research approach is based on nature of research problems, researchers' personal experiences and study audience.

Our research paradigm is therefore Social Constructivists/Interpretivism which believes that individuals develop subjective meanings of their experiences which are varied and multiple, leading the researcher to look for the complexity of views and rely as much as possible on the participants' views of the situation being studied (Creswell, 2014).

Qualitative approach involves understanding the meanings of individuals or group of people attach to a particular social issue. Its data is typically collected in the participants setting while the data analysis inductively builds from particulars to general themes. This argument underscore why data was collected from audience interaction with disseminated health information news on X platform "# HealthyNation" of The Daily Nation @DNHealthyNation.

Narrative Technique

Narrative as a technique in qualitative research was used to generate data. Narrative research is a design of inquiry from the humanities in which the researcher studies the lives of individuals and asks one or more individuals to provide stories about their lives (Riessman, 2008). As Juan &

Ong'ondo, (2011) posits, narrative research process is highly personnel, intimate approach to research that demands a high degree of care and sensitivity on the part of the researcher. Recognizing the nature of narrative technique, research participants were assured of their confidentiality and anonymity throughout the study.

Similarly, equality is critical in the researcher – participant relationship (Friedman and Combs, 1996), thus the study participants were empowered to tell the story through listening to their stories on their perceptions and feelings and understanding of scientific health reports disseminated on X platform "HealthyNation" of the Daily Nation @DNHealthyNation. Some of which are directly and indirectly related to their current health status. The research followed the participants' leads and listened to their stories without any interruptions. Multiple data sources from key information, journals and informal conversations were used to validate the accuracy of the stories.

Study Population and Sample

A population is the group that research focuses on (Cooper and Schindler, 2003). The study targeted three (3) practicing journalists, two (2) medical science researchers and one (1) social science researcher with bias in health communication who acted as key informants. It further targeted fourteen (14) active X users who were following either Nation Media Group X page #NationAfrica "@NationAfrica or The Daily Nations Health and Science Pull Out page #HealthyNation "@DNHealthyNation. Five (5) X users were following @DNHealthyNation X handle while remaining five (5) were following @NationAfrica X handle.

It is important to note that research participants who were following only @NationAfrica X handle were also aware of @DNHealthyNation. This therefore put the study target population at twenty (20). Below is the summary of target audience.

Table 1. Summary of key informants

Name	Profession	Gender
Participant A	Practicing Journalist	Male
Participant B	Practicing Journalist	Female
Participant C	Practicing Journalist	Male
Participant D	Medical Researcher	Female
Participant E	Medical Researcher	Female
Participant F	Social Science Researcher (Specialization: Health Communication)	Male

Table 2. Summary of active X users

No.	Name	X Account Followed	Gender	Age (Yrs)
1	Participant A	@DNHealthyNation	Male	28
2	Participant B	@DNHealthyNation	Male	30
3	Participant C	@DNHealthyNation	Female	30
4	Participant D	@DNHealthyNation	Female	29
5	Participant E	@DNHealthyNation	Female	33
6	Participant F	@NationAfrica	Male	35
7	Participant G	@NationAfrica	Male	26
8	Participant H	@NationAfrica	Male	27
9	Participant I	@NationAfrica	Female	30
10	Participant J	@NationAfrica	Female	34

Convenient sampling was used. Serem, Boit, and Wanyama (2013), posits that this approach involves using whichever individual (elements) are available to the researcher on a first come first served basis. Convenient sampling eased accessibility of the target audience where an active X user who either followed #NationAfrica @NationAfrica OR #HealthyNation @DNHealthyNation was interviewed.

Data Collection Methods

Data was collected through open-ended interviews, desk research, observation of audience interaction on Daily Nation X platform through text and image analysis i.e *Tweets*, *Retweets*, *Likes*, *Shares* and *Replies*. DiscoverText application tool was used to search for Tweets.

Data Analysis

Thematic analysis as a qualitative data analysis strategy was used, resulting to an inductive approach where themes emerged from the data. Thematic analysis is the search for themes of relevance to the research topic under which reasonably large amounts of data from different sources such as observations, interviews and documents can be organized (Hammersley et al., 2001). During data analysis, the following stages were followed: transcribing data, re-familiarising with the data, first phase coding, second phase coding, third phase coding and product report. Data was presented inform of narratives, table and figures.

DATA ANALYSIS AND INTERPRETATION

General Information

Involved as participants in the study were Two medical science researchers' and one social science researcher with a bias in health communication. They helped to answer the following research questions; Are health reports on X framed the same way as they appear on original reports?

How does X audience react to health news reports disseminated on X platform @DNHealthyNation? What are expectations of researchers on disseminated scientific health reports on X.

The sample size of the study was 20 participants. All participants responded in the study thereby making response rate to be excellent at 100 percent. This correlates with Mugenda, (2008) recommendation that a response rate of 50 percent is adequate for analysis and reporting; a rate of 60 percent is good and a response rate of 70 percent and over is excellent.

Desk Research

Desktop research was conducted throughout pre-field, data collection, research analysis and compilation. This included literature review and online sources. The desktop research acted as a basis for understanding different contexts of interactions on social network sites i.e. social, political, religious, academic and economic contexts. Desk research also helps in understanding various social network sites management tools such as TweetDeck which is a powerful tool for real time tracking, organizing and scheduling of Tweets. DiscoverText which helps in searching and analysis of various tweets. SocialBro which is an advanced solution for the management and analysis of X communities. It helps media professionals to do in-depth analysis of their network community.

Interviews

Interviews are one of the most important sources of data in qualitative research, since they allow participants themselves to report their thoughts and experiences thereby giving deeper meanings in understanding social phenomena (Juan & Ong'ondo 2011).

The study adopted open-ended interviews so as to allow deeper exploration of responses by participants, hence probing and exploring emerging dimensions that may not have been previously considered pertinent aspects of this study.

Open-Ended Interviews with Key Informants

Even though Key Informant Interviews (KIIs) have been noted to have a limited scope and tendency towards "biasness" either in selection of respondent or respondents' views, they have greater advantages which include the opportunity to tap into the deep knowledge of respondents (Mwita, Abraham, & Joy, 2021). This advantage was relevant to this study since it wanted a deeper understanding of X users' interaction with disseminated scientific health information and meanings attached to search reactions.

Six Key Informant interviews were conducted to seek more information on how health scientific findings are disseminated on X platform by journalists and expectations of science researchers from such disseminations. Of these informants, three were media practitioners (practicing journalists) and remaining three were medical science researchers and a social science researcher with bias in health communication. Three key informants were men and three where women hence gender balance were factored in the research. Convenience sampling was used for ease of access for the key informants.

Observation

Observation is one of the key data generation methods in qualitative research. It means getting data through critically watching activities of a person with a view of obtaining deeper understanding of activities under study (Cohen et al.,2000).

Observation can be classified in terms of being structured and semi-structured. Semi-structured interviews involve paying attention to the whole event as it takes place and taking field notes, while structured observation involves a pre-designed list of activities that the researcher is going to watch out for in the field (Juan & Ong'ondo, 2011).

Angrosin (2005), posits that there are three observational procedures usually employed in qualitative research, namely, descriptive, focused and selective observation. Where the first one involves trying to capture and note all details, while second concentrates only on events relevant to the topic under investigation, and the third attends to be more particular on categories of the issue of interest.

This study adopted focused observation approach since the interest was in Tweets, Re-tweets, Likes and Shares related to scientific health report disseminations so as to get insights on how X audience react to such information through @DNHealthyNation. The study was therefore not interested on any other Tweets on @DNHealthyNation platform.

Results

The following results were given by the study participants in regard to the research questions that were asked.

Open Ended Interview with Journalists

Research question: Are health reports on X framed the same way as they appear in original reports?
Journalist A: According to the study participant, X platform has a character limitation 280 on its page. This means that one has to use short sentences to relay information to audience. This leads to shortening of sentences which sometimes leads to leaving out a number of key words in research finding that would have clearly communicated the scientific finding on the page. This means that it is only through the links bitly.is link management platform that the reporter is able to share more about the report.

"X platform has a burst of information and the audience are spoilt of choice hence few people will take their time to open the links and read health reports. This means that if the tweeted scientific outcome word phrase is misleading then chances are the audience will also be misled on the disseminated scientific outcome" said the journalist.

Since X has been adopted in the practice of journalism, the journalistic style of writing dominates a lot of Tweets. Journalists are meant to follow rules that involve writing story leads, that is having the 5 Ws and H in mind (Who, where, what, when, why and how). A good lead should not go beyond 21 words according to journalist A. This practice influences journalists to use their own words when framing health scientific disseminations on X platforms which erode the original meaning of the content as intended by the researcher.

For example, on June 2022, a tweet on #HealthyNation page *"Do men have a biological clock? The best age to become a father.* Bit.ly/3OsOcmA. The research was published in the Journal of Epidemiology and Community Health. What is tweeted is journalistic style of writing that depicts a question lead, however exact study title has not been used.

In addition, when you follow provided link; the specific journal where the study is published is stated, however the title of the study isn't mentioned. During the in; *"As you can see there are some erosion that is taking place from the original content"*

Journalist B: *Researchers are not actively engaging with media organizations when disseminating their research findings. In most cases once research studies are sent to journals and once published that becomes the end of the study. It is journalists who takes the initiative to scan through the journals and pick up studies to disseminate to the public. This means journalists are left to interpret key findings to the public without the researcher guidance. This sometimes leads to cases of misinterpretation. The journalist said; "Active engagement of researchers to the media means that, they should send their research findings to media outlets and provide interpretation on key findings. Remember there are specific scientific jargons which only the researcher is able to simplify without losing meaning, if researchers can do this then the meaning of research content will be kept intact. Remember journalists are supposed to simplify and interpret information for ease of public consumption hence they are likely to avoid key medical jargons which are very pertinent in a study"*

Journalist C: *Researchers are yet to fully embrace the power of social network sites in disseminating findings. In most cases researchers believe that what is published in the Daily Nation hard copy newspaper has more impact that information disseminated on X page for the Daily Nations' Health and Science pullout page. In most cases scientists are interested in research findings to be published on traditional print media than online. "You will find that a research publication is disseminated on both print and X platforms, however, the principle investigators who are supposed to promote the study are not even retweeting the study, liking or even sharing on X platform, some don't even follow the media outlet on X platform" Said the Journalist*

In addition, social network sites have tools used in interacting with the public that keep on changing. Journalists should stay up to date with technological changes on these platforms. Most journalists during scientific research disseminations struggle in creating tweets that spark conversations and keep audience engaged. This leads to natural death of tweets hence inadequate reaction of audience with health report disseminations.

Also, journalists need to learn more about social media management tools that can help them learn about content management and understanding various networked/online communities.

Open–Ended Interviews with Researchers

> **Research question:** What are expectations of researchers on disseminated scientific health reports on X?

Since there is character limitation on X platform, journalists often provide links where further readings on a scientific outcome. The pieces provided are journalistic writings which are interpreting the public research findings. It is during this interpretation that the message is twisted. To avoid this, journalists should provide further links where one can get original research so as to help in curbing any misinformation or disinformation that might have occurred during the interpretation.

Competition for readers' attention by journalists often leads to more sensationalized scientific research interpretation by journalists. This twists the original objectives of research findings. Journalists should therefore not use sensationalized language in reporting scientific research findings.

Methods of gathering research disseminations by journalists. While there are different forms of news gathering i.e. telephone conversations, Observations, and interviews. Most journalists do desktop research to source for scientific disseminations. During this process journalists' source and interpret for themselves the key research findings without follow up with the lead investigator. This leads to misinterpretation of key research findings.

Open-Ended Interviews with Active X Users

Research question: How does X audience react to health news reports disseminated on X platform @DNHealthyNation?

Active user C reported that she is mostly attracted to scientific health disseminations that talks about women health issues. On such disseminations she goes through the Tweets, open extra links given to read more about research findings. She further goes ahead to "Like, Share, Comment and Retweet" such information. However, on issues affecting male gender she just read the Tweet and move to other information on the platform. Participants E and G who happens to be of female gender gave the same response.

Participants A and B responded that while scientific health disseminations are very important, they only read the tweets but do not go further to the provided links to read more about research findings, they therefore become passive consumers of such disseminations.

Out of ten (10) study participants, four (4) said to have shared scientific health disseminations from X platforms to other popular social network sites such as Facebook, WhatsApp and Instagram.

When asked if the participants understand the messaging of health reports six (6) participants report that journalists simplify scientific disseminations in language they understand, however four (4) study participants said that they are not sure if journalists are accurately reporting what is contained in the original research findings since they are unable to get links for the original published research.

Study participants also noted that in some instances journalists only tweet but don't provide links for further reading of the articles. This makes health reports to be treated as Fake News by Tweeter users.

Observations

There are various ways in which one can observe how X audience interact, interact and engage with communication on health information i.e through “Reply/Comment, Retweet, Like, Share”.

According to observations on Daily Nations Health and Science pull out X platform @DNHealthyNation there were limited reactions, interactions/engagement with health reports by X users.

Example 1; On 21/06/2022 a tweet on HealthyNation @DN-HealthyNation

“The big salt lie”

“How much salt is too much salt? We dissect the myths and facts about sodium chloride intake”

The tweet had; “No Reply, 1 Retweet, No Like, and No Share”

Example 2: A retweet from @NationAfrica 21/06/2022

“Vaginal discharge: What’s normal and what’s not?”

The tweet had ;” No Reply,5 Retweet, 15 Likes, No Share”

This was perhaps the tweet the highest reaction in the month of June 2022 on @DNHealthyNation page

Example 3: A tweet on @DNHealthNation 9/06/2022

“Scientists explain why old people become weak”

The tweet had; “No Reply, 1 Retweet, No Likes, No Share”

Example 4: A retweet on @NationAfrica May 25/2022

“Large number of malaria mosquitoes bite during the day, shows study”

The tweet had; “No Reply, 3 Retweet, 17 Likes, No Share”

Example 5: Tweet on @DNHealthyNation 5/05/2022

"New insulin tech a relief for Type 2 diabetes patients"
The tweet had; 'No Reply, 2 Retweet,3 Likes, No Share"

Example 6: Tweet @DNHealthyNation 19/04/2022

"Unvaccinated pregnant women at risk of stillbirths, study shows"
The tweet had; "No Reply, No Retweet, No Likes, No Share"

RESEARCH FINDINGS

Researchers and practicing journalists were the key informants in the study. The study established that health reports on most occasions are framed differently from original reports. This was attributed to several factors:

The character limit of 280 on X platform sometimes forces journalists to use short sentences which likely leads to the omission of key words from original research; hence messages get twisted.

The study found that competition for readers' attention by journalists often leads to more sensationalized scientific research interpretation by journalists which changes the messaging. There is also improper follow up of lead research investigators by journalists so that they can interpret research findings. The culture of desktop research when it comes to sourcing scientific health research disseminations by journalists has led to poor interpretation of research findings since journalists search for publications on the internet and do their own interpretation to the public. Inadequate personnel who understand health sciences is also a contributor to poor interpretation of scientific health research disseminations.

Study found out that sometimes they do not actively work with the media, hence once research findings are published, they leave interpretation work to journalists. The study also found out that researchers have not fully embraced the power of social network sites, hence they are only concerned with publication of research findings on traditional print media. It most cases it is the journalists who push scientific disseminations while scientists take back seat. In-fact some breakthrough health research published on @DNHealthyNation platform has no single "Like, Share, or Retweet."

The study found out that there is minimal interaction of health reports by X audience. Men are passive consumers of health disseminations on X platform while most women engage only with content that talks about women health.

Research expectations are sometimes not met when journalists interpret and simplify scientific research findings for ease of public consumption since key jargons that are important in a study are left out, this causes misinterpretation of research findings.

Researchers expect journalists to provide links to original research publications so as to act as a reference in case of any misinformation and disinformation during interpretations. This, however, sometimes isn't provided but instead journalists provide links to their own interpreted journalistic piece.

In addition, journalists need to expose themselves more to social network sites management tools such as "*TweetDeck*" for tweet scheduling, "*Discovertext*" which helps in searching content across multiple Tweeter pages and "SocialBro" which help in understanding and managing various X communities. There are other forms of communication on X platform which can help journalists to interact more with their audience such as Likebaiting which popular way for users on social media to ask for "Likes".

RECOMMENDATION FOR FURTHER RESEARCH

Since this study was limited only to X platform, there is need to do more research on consumption of scientific health disseminations on other social network platforms.

REFERENCES

Aggarwal, C. (2011). *Social Network Data Analytics.* USA Kluwer Academic Publishers Boston/Dordrecht/London. http://charuaggarwal.net/socialtoc.pdf

Austin, J. L. (1975). *How to do things with words.* Oxford at the Clarendon Press. https://www.thoughtco.com/speech-act-theory-1691986

Bormann, E. G. (1985). Symbolic Convergence Theory: A Communication Formulation. *Journal of Communication, 35*(4), 128–138. DOI: 10.1111/j.1460-2466.1985.tb02977.x

Cohen, L.. (2000). *Research Methods in Education* (5th ed.). Routledge Falmer.

Creswell, J. (2014). *Research Design; Qualitative, Quantitative and Mixed Methods Approached* (4th ed.). SAGE.

Donah, M., & Nicole, B. (2007) Social Network Sites: Definition, History, and Scholarship. *Journal of Computer- Mediated Communication.* https://onlinelibrary.wiley.com/doi/full/10.1111/j.1083-6101.2007.00393.x

Endres, Thomas G. (September 1994). Co-existing master analogues in symbolic convergence theory: The knights of Columbus quincentennial campaign. *Communication Studies.* 45 (3–4): 294–308. doi:. ProQuest 233195476DOI: 10.1080/10510979409368430

Friedman, J., & Combs, G. (1996). *Narrative Therapy: The Social Construction of Preferred Realities.* Norton.

Heldman, A. B., Schindelar, J., & Weaver, J. B. (2013). Social Media Engagement and Public Health Communication: Implications for Public Health Organizations Being Truly "Social.". *Public Health Reviews, 35*(1), 1–18. DOI: 10.1007/bf03391698

McBride, K. (2012, October 31). *Social-networking sites in foreign language classes: Opportunities for re-creation.* Academia.edu. https://www.academia.edu/2072765/Social_networking_sites_in_foreign_language_classes_Opportunities_for_re_creation

Mugenda, A. G. (2008). *Social Science Research*, Theory and Principles. *Nairobi: Applied Research &Training Services*. Sciepub.com. https://www.sciepub.com/reference/259964

Mwita, C., Abraham, M., & Joy, C. (2021) *The Kenya Assessments 2021*. Internerws. https://internews.org/wp-content/uploads/2021/04/KMAReport_Final_20210325.pdf

Nordquist, R. (2024). *Speech-Act Theory: Definition and Examples*. ThoughtCo. https://www.thoughtco.com/speech-act-theory-1691986

Portland Communications. (2016). *How Africa Tweets 2016*. Portland. https://portland-communications.com/publications/how-africa-tweets-2016/

Poutintsev, F. (2023, March 4). *13 Advantages and Disadvantages of Social Networking Sites - EU-Vietnam Business Network (EVBN)*. EU-Vietnam Business Network (EVBN). https://evbn.org/13-advantages-and-disadvantages-of-social-networking-sites-1677909672/

Searle, J. R. (1969). *Speech Acts*. Cambridge University Press., https://scholar.google.com/scholar?q=Searle+J.R.+1969:+Speech+Acts&hl=en&as_sdt=0&as_vis=1&oi=scholart

Serem, D. K., Boit, J. M., & Wanyama, M. N. (2013). *Understanding Research; A Simplified Form*. Utafiti Foundation.

Shields, D. C. (2008). *Symbolic Convergence Theory*. Wiley Online Library., DOI: 10.1002/9781405186407.wbiecs125

SIMElab Africa. (2019). *Social Media Consumption in Kenya*: Practices and Trends. Kenya; USIUAfrica. https://www.usiu.ac.ke/assets/file/SIMElab_Social_Media_Consumption_in_Kenya_report.pdf

Chapter 6
Fake News for Whom? Discussing Types, Political Appropriation, and Consumption of False Contents

Ruben Arnoldo Gonzalez
https://orcid.org/0000-0002-6758-5328
Benemérita Universidad Autónoma de Puebla, Mexico

ABSTRACT

Fake news is a problematic concept with multiple meanings for diverse persons. Journalists, politicians, academics, and even laypeople use it on a daily basis. However, more often than not they refer to different things. Therefore, in order to elaborate a more robust analysis of the term, it is important to clarify who utilises it and for what. Hence, the aim of this chapter is threefold: firstly, considering its content, this study proposes an updated typology of fake news that integrates humorous messages, disinformation, and unethical journalism practices. Secondly, it establishes a distinction between the message as such and its weaponization by populist political leaders. Thirdly, it reflects on the reasons why the audience constantly look for - and share - this kind of stories. In doing so, it draws on a comprehensive literature review.

DOI: 10.4018/978-1-6684-9356-4.ch006

INTRODUCTION

What are we talking about when we talk about fake news? Fake news is a problematic concept. It is difficult to define due to the excessive use by media, politicians, scholars, social media users, and even regular people (Freelon & Wells, 2020; Baptista & Gradim, 2021). Thus, it has become a sort of umbrella term utilised by different actors to describe a diverse set of supposedly false content (Farkas & Schou, 2018; Molina et al, 2021; Carson & Wright, 2022). As a result, this buzzword can be considered a hallmark of contemporary political communication (Tandoc Jr. et al, 2018; Molina et al, 2021; Freelon & Wells, 2020). Nonetheless, it merely offers a simplistic understanding of journalism, politics, and digital communication technologies (Camargo & Simon, 2022). Furthermore, its constant presence in popular conversations, political debates, media content, and academic research has fostered a largely exaggerated sense of informational moral panic (Lazer et al, 2018; Wahutu, 2019; Cabañes, 2019; Paolucci, 2023; Murphy, 2023).

This is precisely the reason why fake news is a catch-all term, with multiple definitions (Farkas & Schou, 2018; Mutahi & Kimari, 2020). Having conflicting and ambiguous meanings poses evident conceptual problems, which hinder its correct interpretation (Mutahi & Kimari, 2020; George et al, 2021; Camargo & Simon, 2022; Carson & Wright, 2022). In that sense, Habgood-Coote (2019: 1034) argues that fake news is a "linguistically defective" term, because it does not have a stable meaning and, hence, it is not always clear what it refers to. Therefore, "when a word or phrase can be applied to anything, it means nothing" (McNair, 2017).

It is worth stressing that fake news is not exactly a new phenomenon, because the manipulation of information has a very long history around the world (Allcott and Gentzkow, 2017; Lazer et al, 2018; Bakir & McStay, 2018; Tandoc Jr. et al, 2018 and 2019; Humprecht, 2019; Panagopoulos & Roozenbeek, 2020; Freelon & Wells, 2020; Giusti & Piras, 2020; Baptista & Gradim, 2021; Carson & Wright, 2022). Consequently, the origins of this kind of messages can be tracked down to the development of the writing system, when falsehoods could be registered and kept for longer periods of time (Tandoc Jr. et al, 2018; Fallon, 2019). The arrival of the *penny press* in the XIX century witnessed famous cases of fabrication of news, such as *The Moon Hoax* (Carson & Wright, 2022), or the practice of the *yellow journalism* for business purposes (Molina et al, 2021). The first half of the XX century was also full of cases of false or misleading media content, for instance the famous radio broadcast of *The War of the Worlds* by Orson Wells (Tandoc Jr. et al, 2018 and 2019), or the Nazi and Soviet propaganda (Fallis & Mathiesen, 2019; George et al, 2021).

However, the current use of the concept of fake news has been widely connected with the 2016 US presidential election, when the then-candidate Donald Trump - and to a lesser extent Hillary Clinton – continuously accused political opponents and critical journalist of spreading lies (Bakir & McStay, 2018; Tandoc Jr. et al, 2018; Wahutu, 2019; Molina et al, 2021; Fallon, 2019; Giusti & Piras, 2020; Freelon & Wells, 2020). Another important event that - during the same year - also favoured the continuous diffusion of falsehoods with a political aim was the so-called *Brexit*, because the campaigners for the UK leave from the European Union were suspects of using disinformation for manipulating voters (Humprecht, 2019; Murphy, 2023).

Although it is an old phenomenon, it is safe to claim that digital communication technologies, particularly social media, have facilitated the rapid expansion of fake news (Bakir & McStay, 2018; Humprecht, 2019; Fallon, 2019; Giusti & Piras, 2020; Munoriyarwa & Chambwera, 2020; Galletta Horner et al, 2021). The massive worldwide use of these platforms has turned them into amplifiers of all kinds of messages, and the viralization of fake news in particular (Tandoc Jr. et al, 2018; Freelon & Wells, 2020; Giusti & Piras, 2020; Galletta Horner et al, 2021).

Digital technology promotes user-generated content that competes against mainstream media to attract audiences' attention and traffic (Bakir & McStay, 2018; Tandoc Jr. et al, 2018; Fallon, 2019; Molina et al, 2021; Freelon & Wells, 2020). As a consequence, online platforms have increasingly become the main source of information for many people (Tandoc Jr. et al, 2018; Molina et al, 2021; Newman et al, 2023). Under these circumstances, social media represent the ideal environment for fake news to freely flourish (van der Linden et al, 2020; Baptista & Gradim, 2021; Galletta Horner et al, 2021).

As it has been outlined in this preamble, fake news is not a mere message-oriented aspect. On the contrary, it is a highly complex phenomenon, which involves – at least – three key aspects: content, appropriation, and consumption. Hence, and drawing on a comprehensive literature review, the aim of this chapter is to establish a discussion on those subjects. Consequently, there is a set of research questions derived from these issues: What is the nature of the message? How do political figures use it? And Why do people consume this kind of messages? In addition, and related to the first topic, this study also proposes a typology of fake news. That is, beyond the mere recount of previous works, this chapter seeks to advance the understanding of this topic by presenting an updated taxonomy.

DEFINING FAKE NEWS

Before analysing the diverse types of fake news, and presenting this chapter's own proposal in the next section, it is necessary to define this term. Although fake news is a problematic issue, it is still possible to establish a common ground for its understanding. In doing so, scholars agree on three fundamental features: facticity, format, and intention.

Regarding the first aspect, as its name indicates, fake news is a deliberately fabricated falsehood presented as truth (Loveless, 2020; Mutahi & Kimari, 2020). That is, it represents a fictitious account of events or, at least, a verifiable false content (Allcott & Gentzkow, 2017; Tandoc Jr. et al, 2018; Farkas & Schou, 2018; George et al, 2021). In short, it is a hoax (Giusti & Piras, 2020). Notwithstanding, fake news may not be entirely false, because it may present a calculated meld of both falsehoods and bits of actual facts related to current affairs (Loveless, 2020; Mutahi & Kimari, 2020; Baptista & Gradim, 2021). In other words, beyond elaborated and/or blatant lies, there might frequently be a degree of fallacies as well (Tandoc Jr. et al, 2018; Bakir & McStay, 2018).

The second feature of fake news entails its format. It means that this kind of messages are imitations of real news (Baptista & Gradim, 2021). This is because the former are carefully presented as regular journalistic content, but they do not share the editorial and ethical standards associated with professional media and journalists (Fallis & Mathiesen, 2019;

Loveless, 2020; Kwanda & Lin, 2020; Mutahi & Kimari, 2020; George et al, 2021; Murphy, 2023; Billard & Moran, 2023).

Thirdly, the last aspect has to do with the intention behind fake news. In this regard, this kind of content offers purposely manipulated information (Giusti & Piras, 2020; Baptista & Gradim, 2021). Hence, it has the deliberate aim to mislead and/or deceive the audience, which – due to its generally poor media literacy – it usually cannot distinguish between real and fabricated news stories (Kwanda & Lin, 2020; Baptista & Gradim, 2021; George et al, 2021). In the long run, as it will be discussed in the following pages, it might have serious implications for society and democracy.

Beyond its content, another angle of fake news is related to its viral potential. Namely, the capacity to attract attention and, as a result, to boost the amplification of the message. Furthermore, viralization of fake news is directly connected with emotions (Bakir & McStay, 2018; Galletta Horner et al, 2021). Emotionally-charged stories enhance engagement, because people are easily compelled to interact with them. By massively sharing and commenting, fake news facilitates waves of "emotional contagion" (Bakir & McStay, 2018; Galletta Horner et al, 2021).

Viralization is also the outcome of the increasing ubiquity of digital platforms, particularly social media (Molina et al, 2021; Giusti & Piras, 2020; Kwanda & Lin, 2020; Munoriyarwa & Chambwera, 2020). Besides its emotional accent that promotes engagement, online platforms facilitate the extensive replication of fake news. This is possible because these sites are easy to create but hard to track their creators, they are event-driven, and ephemeral too (Munoriyarwa & Chambwera, 2020).

To sum up, fake news can be defined as a totally – or at least partially – counterfeit and emotionally-charged message, intentionally produced for deceiving its audience, and presented as a legitimate news story, which is potentially viral on social media and other online platforms.

The academic literature consistently suggests that fake news has two main goals, not necessarily mutually exclusive: ideological and economic. The former refers to the political motivation of producing fake news, which is to gain influence within a specific context. Consequently, this type of content aims to support an ideology, a party, a faction, a candidate, and/

or to attack the opponents (Allcott and Gentzkow, 2017; Tandoc Jr. et al, 2018; Humprecht, 2019; Giusti & Piras, 2020; Baptista & Gradim, 2021; Tandoc Jr., 2021; Carson & Wright, 2022). In doing so, this hyper-partisan message (Tuters, 2020) promotes the construction and/or destruction of political identities, by legitimising "us" to the detriment of "them" (Farkas & Schou, 2018). As it will shortly be commented, this angle is close to the concept of propaganda, but there is an important difference.

On the other hand, the second goal means that fake news is a profit-oriented message, because it seeks financial gains (Humprecht, 2019; Loveless, 2020; Tandoc Jr., 2021; Carson & Wright, 2022). Thanks to its mainly online nature, it entails a relatively cheap production with potentially massive revenues (Allcott & Gentzkow, 2017; Tandoc Jr. et al, 2018; Baptista & Gradim, 2021). The strategy for viralization relies on the clickbait logic, focussed on increasing the visits, views, shares, and comments (Allcott & Gentzkow, 2017; Tandoc Jr. et al, 2018; Galletta Horner et al, 2021).

Finally, fake news is frequently associated with the concepts of disinformation and misinformation. The first one is close to propaganda, but it should not be taken as synonym. Propaganda has been typically associated with pejorative notions of mind control, distortion of reality, and mass manipulation (see for instance Silverstein, 1987; Herman & Chomsky, 1988; Chomsky, 1992). Nonetheless, recent studies advocate for a more neutral approach of the concept, considering it as an instrument of governance (e.g. Auerbach and Castronovo, 2013; Rawnsley, 2013). Thus, political actors are constantly publicising their levels of performance and delivery, because that may promote popular support. Conversely, from the producer's side, fake news is part of a disinformation strategy that aims to deliberately deceive and manipulate people (Tumber & Waisbord, 2021; Baptista & Gradim, 2021; Murphy, 2023). In other words, whereas both of them pursue to influence receivers, propaganda is not supposed to utilise deceptive messages (Rawnsley, 2013; Baines, 2014; Zollman, 2019), whilst disinformation is a purposeful false content (Tandoc Jr. et al, 2018; Lazer et al, 2018; Humprecht, 2019; Munoriyarwa & Chambwera, 2020; Giusti & Piras, 2020; Tandoc Jr., 2021). In short, the differentiating key here is facticity and truthfulness. That is, propaganda is related to a formal and institutional political communication strategy, employed for publicising achievements and points of view. On the contrary, rather to keep society informed about – for instance – government's activities, fake news is produced for disinforming audiences.

The second concept linked to fake news is misinformation. Considering the receiver's position, this term represents the unintentional sharing of false information. It happens when audiences with limited media literacy uncritically propagate counterfeit stories. This situation hinders the correct understanding of public issues and, thus, may obstruct democratic development (Tandoc Jr. et al, 2018; Lazer et al, 2018; Munoriyarwa & Chambwera, 2020; Tumber & Waisbord, 2021; Murphy, 2023).

TOWARDS A TYPOLOGY OF FAKE NEWS

Considering that fake news has multiple meanings for different actors, establishing a typology is not a simple task. Even though most of the scholars cited here acknowledge the existence of diverse kinds of fake news (e.g. Allcott & Gentzkow, 2017; Bakir & McStay, 2018; George et al, 2021), to this chapter author's knowledge there are only few actual typologies (Tandoc Jr. et al, 2018; Fallis & Mathiesen, 2019; Sundar et al, 2019). The latter are the foundations for the proposal that will be presented in this section.

Before revising previous taxonomies, it is important to mention that the majority of academics agree on three basic types of fake news: satire, propaganda, and advertising masquerade as news. The first one refers to humorous content related to current affairs and, although it might use certain loose features of real news formats, its aim is to merely amuse the audience (Borden & Tew, 2007; Reilly, 2012; Balmas, 2012; Allcott & Gentzkow, 2017; Bakir & McStay, 2018; George et al, 2021). Regarding propaganda, it still seems to be associated with malevolent activities of mind control and mass manipulation, instead of a regular political communication strategy. Nevertheless, several authors consider it to be fake news (e.g. Tandoc Jr. et al, 2018 and 2019; George et al, 2021). Finally, there are some unethical advertising formulas that seek to deceive potential costumers by presenting the information about any given product as a news story (Tandoc Jr. et al, 2018 and 2019; Sundar et al, 2019; Fallis & Mathiesen, 2019; George et al, 2021).

Based upon a review of research papers, Tandoc Jr. et al (2018) classified the scholars' use of the fake news concept in six categories: news satire (entertainment content inspired by true events); news parody (entertainment fictitious content); fabrication (false or unverifiable content aimed to disinform); photo manipulation (unethical alteration of images also aimed to disinform); advertising and public relations (commercial paid content produced for selling or promoting a product or service, but presented as a news story); and propaganda (ideological or political disinformation).

For their part, Fallis & Mathiesen (2019) elaborated five types of fake news: false news (overt and blatant lies that circulate online, and they are so unreal that they are beyond belief); intentionally deceptive news (fabricated content aimed to make people believe false information); bullshit news (exaggerated information that seeks to bolster up someone's or something's public image); news that lacks truthfulness (harmless false messages that do not intend to change people's minds in any way); and counterfeit news (fabricated content presented as genuine news, which aims to mislead the audience).

Finally, Sundar et al (2019) proposed eight types of online content for the purpose of algorithm-based detection of fake news: real news (average journalistic content produced by quality news organizations); false news (malicious and unverifiable messages that are not factual); polarized and sensationalist content (stories that may not be completely false, but emphasise a particular ideology or point of view through inflammatory discourses); satire (humorous content presented as news in order to openly mock someone or something); misreporting (unintentional false stories published by media, as an outcome of mistakes occurred during the production process); commentary, opinion, and feature writing (op-ed texts that transcend the analysis of current affairs by blurring the limit between information and opinion, or by making mere assertions); persuasive information (advertisements and promotional content); and citizen journalism (unverified and inaccurate information, frequently mixed with personal commentary, posted online by social media users).

Drawing on these three highly influential works, the final part of this section presents this chapter author's own proposal of fake news typology (see Figure 1), which purpose is to condense and refine previous attempts. This updated classification includes three broad categories: humour, disinformation, and unethical journalism.

As the first category of the typology, Humour is integrated by two kinds of fake news: satire and parody. Both of them are openly entertainment-oriented messages but, whereas the former is somewhat associated with actual events, the latter tends to be mainly fictitious content (Tandoc Jr. et al, 2018). Rather than to inform the audience, this type of fake news aims to merely be funny. Hence, there is a tacit agreement between sender and receiver that these messages are not supposed to be taken seriously. By loosely using certain characteristics of real news, they consistently scorn public figures' sayings and doings[1] (Borden & Tew, 2007; Reilly, 2012; Balmas, 2012; Allcott & Gentzkow, 2017; Bakir & McStay, 2018; George et al, 2021).

Disinformation is the second category of this proposal, and it is comprised of ideological and commercial contents. This form of fake news seeks to deceive people and, in doing so, they deliberately contain false information. Therefore, the audience is not expected to realise that it might be manipulated by the producer of these fake news (Tandoc Jr. et al, 2018; Lazer et al, 2018; Humprecht, 2019; Munoriyarwa & Chambwera, 2020; Giusti & Piras, 2020; Tandoc Jr., 2021). Ideological messages are primarily related to politics and religion,[2] whilst commercial have to do with the sales and promotion of products and services.[3] It is worth stressing that these stories should not be considered as propaganda and advertising, because these types of professional communication are expected to be carried out following the highest ethical standards. On the contrary, fake news creators do not adhere to any professional norm.

The last category of this classification is Unethical Journalism. On the one hand, it includes the unverified reproduction of falsehoods, and – on the other – the fabrication of news stories. It is important to explain that minor reporting mistakes are purposefully excluded from this category, because they are completely unintentional, random, and even expected errors that are part of the job (Allcott & Gentzkow, 2017; McNair, 2017; Mutahi & Kimari, 2020; Tandoc Jr., 2021). Conversely, the unverified reproduction of falsehoods is a major flaw, because journalists are obliged to double-check every fact and piece of information that they publish.[4] However, the ultimate journalistic offense is the fabrication of news stories, because it contravenes every ethical standard of the profession.[5] Just like disinformation, it may be hard for the receiver to realise that he/she is consuming counterfeit content presented as real news.

Figure 1. Typology of fake news

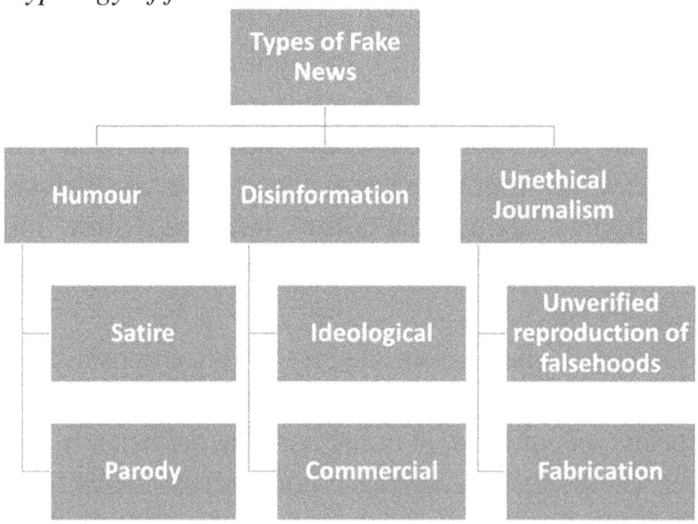

(Own elaboration)

APPROPRIATION BY POLITICAL FIGURES

As it was commented before, fake news is a problematic concept because it has multiple meanings and, hence, it is used in many confusing situations (Tandoc Jr. et al 2018; Molina et al, 2021). On that account, this term is simultaneously a genre and a label (Farkas & Schou, 2018; Tandoc Jr., 2021; Carson & Wright, 2022). That is, it refers both to a kind of message and a tag attached to a content. It is precisely the latter sense that will be discussed in this section.

Considering that most politicians are not used to being criticised, because they only like to be flattered, there is a notorious increase in the political use of fake news as a label. Worldwide politicians have appropriated the term in order to weaponize it against their opponents (Molina et al, 2021; Tandoc Jr., 2021; Carson & Wright, 2022). The aim of these actors is to discredit media and journalists' critical reporting and, in doing so, to delegitimise unfavourable news coverage (Farkas & Schou, 2018; Tandoc Jr. et al 2018; Fallis & Mathiesen, 2019; Tuters, 2020; Freelon & Wells, 2020). However, not only investigative pieces or average news stories are subjected to bear this tag. On the contrary, it applies to any content that contradicts their discourse and opinions (Molina et al, 2021; Reyna, 2024).

As a result of the constant attack by political figures, credibility of the press is likely to be affected, particularly in the eyes of the formers' supporters (Gounari, 2018; Kwanda & Lin, 2020; Carson & Wright, 2022). This circumstance reflects the social polarization of contemporary societies, in which opinions are more important than facts. Therefore, the winner of the struggle between politicians and journalists would determine what is true and what is false (Farkas & Schou, 2018; Molina et al, 2021). Even though this tends to be a global phenomenon, politicians with a more defined populist profile are prone to rely on this pejorative discourse towards critical news outlets[6] (Hanitzsch et al, 2018; Nelson & Kim, 2020; Tant, 2021; Reyna, 2024).

Although Donald Trump gave massive salience to fake news as an attempt to dismiss journalists who dare to contradict him[7] (Farkas & Schou, 2018; Gounari, 2018; Fallis & Mathiesen, 2019; Tuters, 2020; Freelon & Wells, 2020), previous US presidents also utilised the same reasoning when referring to unfavourable coverage of their activities. Holzer (2021) recounts a wide array of conflicts between former presidents and the press. For instance, Thomas Jefferson called certain reporters "divulgers of false news"; Abraham Lincoln said that "newspapers lie, and then they re-lie"; Franklin Roosevelt insulted a journalist by saying he was a "chronic liar", or Ronald Reagan accused media of diffusing "false stories".

FAKE NEWS CONSUMPTION

The third key aspect of this wide discussion on fake news is focussed on its consumption and impact. As it has been stressed throughout previous pages, fake news is not a new phenomenon, but it has currently gained massive exposure due to social media and populist politicians.

In order to explain why people constantly consume fake news, it is useful to outline a wide context. Although it is highly contested as well, the concept of post-truth serves as the background where this kind of content proliferates. Post-truth refers to a political phase in which persons tend to accept certain arguments based upon their own feelings and beliefs, rather than upon facts (Giusti & Piras, 2020). That is, despite their truthfulness and facticity, messages are only valid when they coincide with previous assumptions. Under these conditions of irrationalism in public communication, suspicion towards expertise and science increases, and - as a result - the acceptance of fake news grows as well (Giusti & Piras, 2020; Tuters, 2020; Tumber & Waisbord, 2021).

Since facts are debatable, what really matters then is what a person believes or feels to be true, even though it is not (Bakir & McStay, 2018; Tuters, 2020; Loveless, 2020; Mutahi & Kimari, 2020). Thus, the constant search for reaffirmation of what someone thinks facilitates the formation of groups of like-minded individuals, known as echo chambers (Allcott & Gentzkow, 2017). These spaces are homogeneous social networks with limited tolerance to alternative and opposite views, but with a high level of political and social homophily (Lazer et al, 2018; O'Connor & Weatherall, 2019; Cabañes, 2019; Tong et al, 2021; Baptista & Gradim, 2021; Salaverría & Cardoso, 2023). Hence, this psychological bias leads to partisan reasoning, and an "us versus them" mentality that normalises prejudices (van der Linden et al, 2020; Mutahi & Kimari, 2020; Tong et al, 2021). Thus, different types of fake news are produced by specific target audiences, ergo echo chambers (Baptista & Gradim, 2021).

Both legacy and social media offer limited visions of the world, which boosts the creation of echo chambers. In return, the latter promotes the simultaneous polarization of audiences and media (Lazer et al, 2018; van der Linden et al, 2020; Mutahi & Kimari, 2020; Baptista & Gradim, 2021; Tong et al, 2021). Therefore, consumers look for contents that reaffirm their opinions and then, algorithms do the rest (Baptista & Gradim, 2021). In sum, echo chambers are a fertile ground for the spread of fake news (Mutahi & Kimari, 2020; Baptista & Gradim, 2021).

Echo chambers are entwined with filter bubbles. The latter term means the selective exposure of already known and accepted arguments, no matter how truthful they might be (Loveless, 2020). In other words, echo chambers are a sort of intellectual isolation spheres that prevent the access to antagonistic perspectives (Allcott & Gentzkow, 2017; Baptista & Gradim, 2021). Therefore, members of particular audiences feel comfortable within this safe space where they constantly reinvigorate their value system (Baptista & Gradim, 2021). In short, by the restrictive consumption of previously tolerated ideas, filter bubbles facilitate the formation of echo chambers.

Another aspect that favours the consumption of fake news is that they rely on emotional cues (Giusti & Piras, 2020; Baptista & Gradim, 2021). For that reason, they appeal to strong sentiments rather than to intellectual arguments (Loveless, 2020; Baptista & Gradim, 2021). That is, in order to attract attention, this type of messages contain false heated stories. As an "emotional weapon", fake news is used within ideologically similar networks; that is, echo chambers (Loveless, 2020; Tong et al, 2021). Most of the times, the emotions incorporated into counterfeit stories seek to outrage the audience and foster the "us versus them" feeling (Bakir & McStay, 2018).

Increasing access to a vast amount of information may not necessarily be beneficial, because it requires some basic method to curate a wide array of messages (Loveless, 2020; Tandoc Jr. et al, 2018; Munoriyarwa & Chambwera, 2020; Galletta Horner et al, 2021). This is because internet is open to both professional journalists and unscrupulous fake news creators (Giusti & Piras, 2020). Thus, nearly anyone can produce online content, but not everyone knows how to distinguish between false and true information (Tandoc Jr. et al, 2018; Giusti & Piras, 2020). In doing so, certain degree of expertise in handling digital tools is required (Wahutu, 2019; Loveless, 2020; Giusti & Piras, 2020; Cheng & Chen, 2021). Consequently, in terms of media literacy, the less prepared user is an easy prey of fake news producers (Tandoc Jr. et al, 2018 and 2019; Giusti & Piras, 2020).

The lack of time and training to detect fake news facilitates its impact on the audience (Giusti & Piras, 2020). Notwithstanding, compared with consumption, the influence of this kind of content is not as straightforward as it may appear at first glance (Lazer et al, 2018). In other words, it is highly difficult to establish a clear correlation between being exposed to fake news and behavioural changes, at least in political terms (Lazer et al, 2018; Cabañes, 2019; Fallon, 2019; Giusti & Piras, 2020; Mutahi & Kimari, 2020). Consumers of false stories are usually less informed and, thus, ill-prepared for distinguishing legitimate media, demanding quality information, and making rational decisions (Allcott & Gentzkow, 2017; Bakir & McStay, 2018; Tandoc Jr. et al, 2019; Humprecht, 2019; O'Connor & Weatherall, 2019; Mutahi & Kimari, 2020; Munoriyarwa & Chambwera, 2020). However, several scholars suggest that there is indeed a negative impact on political culture in terms of beliefs (e.g. *Flat Earthers* or *Anti-vaxxers*), attitudes (e.g. cynicism, xenophobia), emotions (e.g. hate), and behaviours (e.g. violent demonstrations); thereby delegitimising democracy and its core values (Tandoc Jr. et al, 2019; Humprecht, 2019; Loveless, 2020; Munoriyarwa & Chambwera, 2020; Kwanda & Lin, 2020; Tuters, 2020; Baptista & Gradim, 2021; Salaverría & Cardoso, 2023; Mauk & Grömping, 2023; Billard & Moran, 2023).

CONCLUSIONS

It has been consistently emphasised throughout this chapter that fake news is a complex phenomenon, and a problematic concept. This is because it involves content production, appropriation by politicians, and consumption. Therefore, it is relevant to understand which one of these issues is being discussed, hence, the term could be better interpreted according to its context. For that reason, the phrase adopts a different sense when it refers to a message as such, when a populist leader uses it, or when the audience interacts with the content. Thus, previous to any analysis, it is important to ask "Fake news for whom?". In doing so, this final section is the place for a brief reflection on these three aspects.

Regarding content production, by proposing an improved typology, this chapter contributes to the study of fake news as messages. Although all types of stories included in this classification are false, not everyone has the same goal. In order to determine that, intention and understanding between sender and receiver are the key. On the one hand, whereas humorous fake news only seeks to amuse the audience, disinformation producers openly want to deceive people for either political or commercial gain. The latter might also apply to the unethical fabrication of stories performed by very specific individual journalists (see McNair, 2007). The case of the unverified reproduction of falsehoods may be just an error, but it is an important one. Therefore, more than immediate gain, it would eventually affect both the reporter and the news outlet's reputation. Furthermore, in spite of its lack of malicious intention, it might as well contribute to disinform the audience.

On the other hand, agreement between sender and receiver is fundamental for differentiating false stories. In that sense, humorous fake news is expected to be taken as a joke by the audience, because the producer's goal is not to deceive the public, but to entertain it. In doing so, the former's intention is supposed to be transparent for the latter. That is not the case of disinformation and unethical journalism (particularly the fabrication of news stories), due to the sender's hidden purpose of misleading the receiver. In these cases, the average member of the audience would find it hard to distinguish between real and counterfeit news.

Related to politicians' appropriation of the term, they may or may not know what fake news really means, but they will certainly continue using it as weapon against their opponents. Labelling as fake news anything that contradicts their discourse has proven beneficial to their own interests, hence, there is no reason why stop doing it. Especially as long as it resonates with their supporters. Nonetheless, this strategy may affect the political communication process in terms of decreasing public trust in both journalism and political institutions; but also between journalists and politicians.

With respect to the receivers' side, consumption determines the offering. Thus, on the condition that the audience demands it, producers will supply fake news. Since polarised societies foster filter bubbles and echo chambers, this kind of messages will have the adequate environment to develop then. Even though media literacy could eventually reduce the problem, it is unlikely that the already convinced people would accept that what they consume is not entirely true.

Finally, as any academic endeavour, this chapter has limitations. Considering that it is a conceptual discussion, it requires empirical verification. Firstly, in pursuance of testing its accuracy, this typology must be applied to specific cases. However, before reaching that point, the proposed classification requires a follow-up analytical process in order to operationalise it. Following Otto's et al (2017) notion of "eroding of boundaries", the differences between these forms of false messages demand a more granular approach. Therefore, rather than a fully operational model, this proposal should be considered as the basic coordinates to navigate through the sea of fake news. Secondly, public figures – particularly politicians – continuously use this term, so it might be useful to analyse how, why, and under which circumstances they resort to this label. Thirdly, audience is a highly complex object of study, hence, it is necessary to understand the reasons and habits behind its consumption of fake news (see for instance Cheng & Chen, 2021; Tong et al, 2021). Despite the growing academic literature that reports findings directly from the fieldwork, most of it comes from Western countries. Thus, research from the Global South, where populism is on the rise, is still much needed.

REFERENCES

Allcott, H., & Gentzkow, M. (2017). Social media and fake news in the 2016 election. *The Journal of Economic Perspectives, 31*(2), 211–236. DOI: 10.1257/jep.31.2.211

Bakir, V., & McStay, A. (2018). Fake News and The Economy of Emotions. *Digital Journalism (Abingdon, England), 6*(2), 154–175. DOI: 10.1080/21670811.2017.1345645

Balmas, M. (2012). When fake news becomes real: Combined exposure to multiple news sources and political attitudes of inefficacy, alienation, and cynicism. *Communication Research, 41*(3), 1–25. DOI: 10.1177/0093650212453600

Baptista, J. P. & Gradim, A. (2021). "Brave New World" of fake news: How it works. *Javnost-the public*, 28(4), 426-443. DOI: 10.1080/13183222.2021.1861409

Billard, T. J., & Moran, R. E. (2023). Designing Trust: Design Style, Political Ideology, and Trust in "Fake" News Websites. *Digital Journalism, 11*(3), 519–546. DOI: 10.1080/21670811.2022.2087098

Borden, S. L., & Tew, C. (2007). The role of journalist and the performance of journalism: Ethical lessons from "fake" news (seriously). *Journal of Mass Media Ethics, 22*(4), 300–314. DOI: 10.1080/08900520701583586

Cabañes, J. V. A. (2019). Digital Disinformation and the Imaginative Dimension of Communication. *Journalism & Mass Communication Quarterly, 97*(2), 435–452. DOI: 10.1177/1077699020913799

Camargo, C. Q., & Simon, F. M. (2022). Mis-and disinformation studies are too big to fail: Six suggestions for the field's future. *Harvard Kennedy School Misinformation Review, 3*(5). Advance online publication. DOI: 10.37016/mr-2020-106

Carson, A., & Wright, S. (2022). Fake news and democracy: Definitions, impact and response. *Australian Journal of Political Science, 57*(3), 221–230. DOI: 10.1080/10361146.2022.2122778

Cheng, Y., & Chen, Z. F. (2021). The influence of presumed fake news influence: Examining public support for corporate corrective response, media literacy interventions, and governmental regulation. In Pompper, D., & Hoffman, L. (Eds.), *What IS News?* (pp. 103–127). Routledge.

Fallis, D., & Mathiesen, K. (2019). Fake news is counterfeit news. *Inquiry : a Journal of Medical Care Organization, Provision and Financing*, •••, 1–20. DOI: 10.1080/0020174X.2019.1688179

Fallon, K. (2019). *Where Truth Lies: Digital Culture and Documentary Media after 9/11*. University of California Press.

Farkas, J. & Schou, J. (2018). Fake News as a Floating Signifier: Hegemony, Antagonism and the Politics of Falsehood. *Javnost - The Public*, 25(3), 298-314. DOI: DOI: 10.1080/13183222.2018.1463047

Freelon, D., & Wells, C. (2020). Disinformation as political communication. *Political Communication*, *37*(2), 145–156. DOI: 10.1080/10584609.2020.1723755

Galletta Horner, C., Galletta, D., Crawford, J., & Shirsat, A. (2021). Emotions: The Unexplored Fuel of Fake News on Social Media. *Journal of Management Information Systems*, *38*(4), 1039–1066. DOI: 10.1080/07421222.2021.1990610

George, J., Gerhart, N., & Torres, R. (2021). Uncovering the Truth about Fake News: A Research Model Grounded in Multi-Disciplinary Literature. *Journal of Management Information Systems*, *38*(4), 1067–1094. DOI: 10.1080/07421222.2021.1990608

Giusti, S., & Piras, E. (2020). Introduction: In search of paradigms: Disinformation, fake news, and post-truth politics. In Giusti, S., & Piras, E. (Eds.), *Democracy and fake news: information manipulation and post-truth politics* (pp. 1–16). Routledge.

Gounari, P. (2018). Authoritarianism, Discourse and Social Media: Trump as the 'American Agitator. In Morelock, J. (Ed.), *Critical Theory and Authoritarian Populism* (pp. 207–227). University of Westminster Press.

Habgood-Coote, J. (2019). Stop talking about fake news! *Inquiry : a Journal of Medical Care Organization, Provision and Financing*, *62*, 9–10, 1033–1065. DOI: 10.1080/0020174X.2018.1508363

Hanitzsch, T., Van Dalen, A., & Steindl, N. (2018). Caught in the Nexus: A Comparative and Longitudinal Analysis of Public Trust in the Press. *The International Journal of Press/Politics*, *23*(1), 3–23. DOI: 10.1177/1940161217740695

Humprecht, E. (2019). Where 'fake news' flourishes: A comparison across four Western democracies. *Information Communication and Society*, *22*(13), 1973–1988. DOI: 10.1080/1369118X.2018.1474241

Kwanda, F. A., & Lin, T. T. C. (2020). Fake news practices in Indonesian newsrooms during and after the Palu earthquake: A hierarchy-of-influences approach. *Information Communication and Society*, *23*(6), 849–866. DOI: 10.1080/1369118X.2020.1759669

Lazer, D. M. J.. (2018). The science of fake news. *Science*, *359*(6380), 1094–1096. DOI: 10.1126/science.aao2998

Loveless, M. (2020). Information and democracy: Fake news as an emotional weapon. In Giusti, S., & Piras, E. (Eds.), *Democracy and fake news: information manipulation and post-truth politics* (pp. 64–76). Routledge.

Mauk, M., & Grömping, M. (2023). Online Disinformation Predicts Inaccurate Beliefs About Election Fairness Among Both Winners and Losers. *Comparative Political Studies*, •••, 1–34. DOI: 10.1177/00104140231193008

McNair, B. (2017). Fake news-a user's guide. *The Conversation*, March, 1-6. https://theconversation.com/fake-news-a-users-guide-73428

Molina, M. D., Sundar, S. S., Le, T., & Lee, D. (2021). "Fake News" Is Not Simply False Information: A Concept Explication and Taxonomy of Online Content. *The American Behavioral Scientist*, *65*(2), 180–212. DOI: 10.1177/0002764219878224

Munoriyarwa, A., & Chambwera, C. (2020). Who Are the Arbiters of Truth? Mainstream Journalists' Responses to Fake News during the 2017 Zimbabwe Coup. *Communicatio*, *46*(4), 75–95. DOI: 10.1080/02500167.2020.1854805

Murphy, K. M. (2023). Fake News and the Web of Plausibility. *Social Media + Society*, *9*(2), 1–18. DOI: 10.1177/20563051231170606

Mutahi, P., & Kimari, B. (2020). Fake News and the 2017 Kenyan Elections. *Communicatio*, *46*(4), 31–49. DOI: 10.1080/02500167.2020.1723662

Nelson, J. L., & Kim, S. J. (2020). Improve Trust, Increase Loyalty? Analyzing the Relationship Between News Credibility and Consumption. *Journalism Practice*, *15*(3), 348–365. DOI: 10.1080/17512786.2020.1719874

Newman, N., Fletcher, R., Eddy, K., Robertson, C. T., & Nielsen, R. K. (2023). *Reuters Institute Digital News Report 2023*. Reuters Institute for the Study of Journalism.

O'Connor, C., & Weatherall, J. O. (2019). *The misinformation age: How false beliefs spread*. Yale University Press.

Otto, L., Glogger, I., & Boukes, M. (2017). The softening of journalistic political communication: A comprehensive framework model of sensationalism, soft news, infotainment, and tabloidization. *Communication Theory*, *27*(2), 136–155.

Paolucci, C. (2023). Pre-Truth: Fake News, Semiological Guerrilla Warfare, and Some Other Media and Communication "Revolutions". *Media and Communication*, *11*(2), 101–108. DOI: 10.17645/mac.v11i2.6628

Reilly, I. (2012). Satirical fake news and/as American political discourse. *The Journal of American Culture*, *35*(3), 258–275.

Reyna, V. H. (2024). "A Mercenary, a Thug … not Even a Journalist": The Stigmatization of News Workers in Mexico. *Journalism Practice*. Advance online publication. DOI: 10.1080/17512786.2023.2300275

Salaverría, R. & Cardoso, G. (2023). Future of disinformation studies: emerging research fields. *Profesional De La información Information Professional*, 32(5), 1-7. https://doi.org/DOI: 10.3145/epi.2023.sep.25

Tandoc, E. C.Jr. (2021). Fake news. In Tumber, H., & Waisbord, S. (Eds.), *The Routledge companion to media disinformation and populism* (pp. 110–117). Routledge.

Tandoc, E. C.Jr, Lim, D., & Ling, R. (2019). Diffusion of disinformation: How social media users respond to fake news and why. *Journalism, 21*(3), 381–398. DOI: 10.1177/1464884919868325

Tandoc, E. C.Jr, Lim, Z. W., & Ling, R. (2018). Defining "Fake News": A typology of scholarly definitions. *Digital Journalism (Abingdon, England), 6*(2), 137–153. DOI: 10.1080/21670811.2017.1360143

Tong, C., Gill, H., Li, J., Valenzuela, S., & Rojas, H. (2021). "Fake news is anything they say!"—Conceptualization and weaponization of fake news among the American public. In Pompper, D., & Hoffman, L. (Eds.), *What IS News?* (pp. 153–176). Routledge.

Tumber, H., & Waisbord, S. (Eds.). (2021). *The Routledge companion to media disinformation and populism.* Routledge.

Tuters, M. (2020). Fake news and the Dutch YouTube political debate space. In Rogers, R., & Niederer, S. (Eds.), *The Politics of Social Media Manipulation* (pp. 218–238). Amsterdam University Press.

van der Linden, S., Panagopoulos, C., & Roozenbeek, J. (2020). You are fake news: Political bias in perceptions of fake news. *Media Culture & Society*, *42*(3), 460–470. DOI: 10.1177/0163443720906992

Wahutu, J. S. (2019). Fake News and Journalistic "Rules of the Game". *African Journalism Studies*, *40*(4), 13–26. DOI: 10.1080/23743670.2019.1628794

ENDNOTES

1 American satire and parody content site *The Onion* (https://www.theonion.com/) is full of stories that illustrate both types of humorous fake news.

2 An example of ideological misleading content related to the origins of the COVID outbreak was published by Fox News web site in April 2020 (https://www.foxnews.com/politics/coronavirus-wuhan-lab-china-compete-us-sources), echoing former president Donald Trump's beliefs.

3 In order to catch readers' attention and take them to a fake site pretending to be *El Universal* official online edition (one of the most prestigious Mexican newspapers), a bitcoin firm published a completely fabricated story involving a high profile female journalist, suggesting that she used Bitcode Method platform and earned a lot of money, which was obviously a scam (https://factormob.com/noticias/?fbclid=IwAR3-YuwNQlFpyMSi_g4CZDdlaVVeVLqPR6JehafI_kceDLEW2GGwHQBEebI).

4 At the beginning of Israel's attacks on Gaza, an unverified story about 40 decapitated babies by Hamas militants was uncritically replicated by worldwide news outlets. The story was debunked shortly after (see for instance https://www.aljazeera.com/opinions/2023/10/13/watching-the-watchdogs-babies-and-truth-die-together-in-israel-palestine).

5 Even though the fabrication of news by individual reporters or news organizations is the exception rather than the rule, there are infamous cases such as the story of Stephen Glass, a journalist that worked for the American magazine *The New Republic*, who had to admit that he made up many of his most notable reports (https://newrepublic.com/article/120145/stephen-glass-new-republic-scandal-still-haunts-his-law-career).

6 During his presidential term, Andrés Manuel López Obrador held a daily morning press conference, and every Wednesday, it included a section called "Who is who in the lies of the week". In order to disprove media's performance, the host of the section used fake news as a tag attached to messages that criticised the administration. In doing so, she labelled those contents as "fake", "malicious", or "misleading" (see also Reyna, 2024).

7 One of the many times Donald Trump accused media of spreading fake news was on February 15, 2017, when he twitted "The fake news media is going crazy with their conspiracy theories and blind hatred. @MSNBC & @CNN are unwatchable. @foxandfriends is great!".

Chapter 7
Adoption of Information Communication Technologies in Enhancing Agricultural Productivity in Rural Kenya

Dorine Lugendo
https://orcid.org/0009-0009-1217-7798
Maseno University, Kenya

ABSTRACT

With the development of Information Communication Technologies(ICTs), family engagement and to a larger extent social engagement has evolved from traditional media to digital media. Many countries have shifted from using agricultural extension officers to using of digital technology in accessing agriculture related information to enhance overall economic growth. Farmers in Kenya continue to face a multitude of problems to maximize crop and animal productivity. It is a fact that Kenya possesses valuable agricultural knowledge and expertise in addition to fertile soils.

DOI: 10.4018/978-1-6684-9356-4.ch007

However, a wide information gap exists between research knowledge and the farmers. Farmers need timely expert advice to make them more productive and competitive.

INTRODUCTION

Traditionally, agricultural extension officers operated as intermediaries between research and farmers as they facilitated communication in helping farmers in their decision-making and ensuring that appropriate knowledge is implemented to obtain best results from Agricultural activities. Agricultural extension is one of the institutional sectors that advance the exchange and transfer of information that can be changed into useful knowledge. It fills in as a source of guidance and help for farmers in enhancing their marketing and production Agricultural extension is one of the oldest sub-disciplines of development communication. Servaes & Lie (2020) maintain that the agricultural extension officers play an important role in rural communication in regard to agriculture. These officers are often the first to be in direct contact with rural households, and so these people can signal rural household problems.

In Kenya, Agricultural extension services have been in existence since the establishment of the British Colonial government to offer guidance to farmers in order to boost agricultural production in addressing issues of food security. They provide crucial knowledge, building capacity, and connecting farmers to resources and markets, they also empower rural communities to achieve food security and economic prosperity.

Chelule, Munyua and Kibett (2022) argue that Agricultural extension services are put in place to improve rural development in many nations however their financing has been a major impediment in them accomplishing their mandate. They aver that in Ghana as much as the Department of Agriculture (DOA) is decentralized, it depends on the central government to support its work. Although this money is intended to be distributed quarterly, at the beginning of each quarter, the analysis discovered that they have not been forthcoming since 2012.

A study in Uganda indicates that subventions from the central government are crucial for funding agricultural extension programs in Uganda. Local governments in Uganda became more reliant on central government financing after the graduated tax, which accounts for 80% of local income, was repealed in 2005.

In Kenya, Agricultural extension services have been in existence since the establishment of the British Colonial government to offer guidance to farmers in order to boost agricultural production in addressing issues of food security. Although agricultural services function was devolved to the county, the money for Agricultural Extension Officers did not follow them.

The agricultural sector should employ contemporary methods and technologies so as to modernize agriculture and enhance productivity.

Msoffe & Ngulube (2016) point out that Information and communication technology (ICT) can help in delivering timely and relevant extension services through mobile-enabled agricultural services, weather advisories, etc. Governments should invest in mobile or web-based innovations which are simple, user friendly and cost effective to be used by farmers. These innovations have to be supported by research. In Kenya, food production may not improve unless efforts are taken to modernize the agricultural sector by providing it with modern inputs such as quality seeds of improved varieties, fertilizers and pesticides, better cultivation methods, application of modern tools and farm equipment.

Due to the relatively low quality, low yield agricultural products and the challenges of climate change there is an urgent need to identify possible technological innovations, which could be implemented to bridge this information gap between the research finding and practice. Knowledge and information are major drivers for development in improving levels of agricultural productivity in rural Kenya.

Oyindeinbofa (2018) points out that when rural farmers lack access to knowledge and information necessary to help them achieve maximum agricultural yield they are not only left in the dark but are left with no choice than to move to urban centers in search of formal employment. They add that farmers need to access to knowledge and information in areas of new agricultural technologies, planting treated seeds, soil conservation, prevention of plants and animal diseases, fertilizer application, proper storage of farm products, marketing techniques, co-operative and other agro-cultural activities. However, the advancement in technology ensures provision latest expert advice in a simple, personalized and timely manner to the farmer thereby reducing the knowledge gap between the expert and the farmer. New technologies and innovations can play a central role in improving agriculture productivity and sustainability in Kenya and Africa at large. Hamid, Tarek, Francesco, Sinisa & Roberto (2021) however posit that adoption of innovations and new technologies in agriculture can not only have a central contribution to the achievement of food security and the improvement of the sustainability of agri-food systems if there is deployment of appropriate technologies to address the pressing challenge of climate change. The impacts of these technologies depend on the level of acceptance by farmers/producers and the types of technologies adopted. It is therefore of paramount importance to understand the factors that affect the adoption of technologies in agriculture worldwide and to fine-tune the acquired knowledge to the specific context of each country.

Many countries have shifted from using agricultural extension officers to using of digital technology in accessing agriculture related information to enhance overall economic growth. Digital agricultural services have proliferated across Africa over the last decade. Most are services that work on mobile phones, although more advanced technologies are in use too – like satellite images, sensors, block chains and big data analytics. These services offer access to information, markets and financial products. Information and communication technology (ICT) can help in delivering timely and relevant extension services through mobile-enabled agricultural services and weather advisories.

Abebaw (2020) argues that Information Communication Technologies (ICTs) is one important mechanism and offer a great opportunity to solve the problem of information problem to smallholder farmers. More households in developing countries own a mobile phone than have access to electricity or clean water, and nearly 70%of the bottom fifth of the population in developing countries own a mobile phone. He adds that a mobile phone brings an enormous prospect to disseminate agricultural information through SMS, IVR (Interactive voice response), internet, radio through mobile phone. ICTs facilities exchange of information among farmers from diverse social groups, allowing rapid access to information needed for exchanging, buying, producing, and selling products and lead to increased productivity gain.

Various ICT initiatives in agriculture have been taken up by the Kenyan government, DigiFarm, a mobile phone application offers farmers convenient, one-stop access to quality farm inputs at discounted prices, input loans, learning content on farming as well as access to market for their farm produce. Mobile phone usage by the farmers can reduce the information search costs, thereby dramatically lowering transaction costs and enabling greater participation of the farmers in commercial agriculture. Kukreja and Chakrabarti (2013) conclude that there are convincing reasons that the ICTs are the strongest tools for transfer of agricultural knowledge and technologies to the farmers. However, they insist that there is a need to create awareness among the young and middle-aged famers about the benefits of the services provided by ICT, which thereby will lead to increase in usage of ICTs. They further add that in order to achieve sustainable agricultural development in any country, there is a need to include the ICTs into the structural and functional components of the extension organizations.

LITERATURE ON AGRICULTURE IN KENYA

Agricultural Information Dissemination

Agricultural production depends largely on the availability of and access to relevant information. Information helps to provide opportunities for rural farmers to improve their farming activities, while helping to improve their livelihoods. Therefore, improved systems for dissemination of agricultural information to the rural areas can improve farmers' access to information, which may assist them in making informed decisions about their farming activities. Msoffe and Ngulube (2016) say that effective dissemination of agricultural information plays a key role in supporting rural agricultural activities as it may improve farmers' knowledge and increase their agricultural productivity.

Agricultural information services in rural areas of developing countries have been provided through various channels such as radio, print, video, television, films, pictures, drama, dance, folklore, group discussion meetings, exhibitions and practical demonstrations. However, rural farming communities vary, thus require information services based on models that are context specific.

Msoffe and Ngulube (2016) argue that farmers in the rural communities interact with the information system in an effort to satisfy their information needs. The information system that is designed with a proper understanding of the target community will be in a position to offer appropriate information to satisfy farmers' needs. They add that assessment and understanding of the rural community could be based on, but not limited to the following: information needs, information seeking patterns, farmers' characteristics, preferred information sources, appropriate sources, cultural and personal values, and farmers' competence. Information providers should therefore make it a priority to have adequate knowledge about the target community as a prerequisite for planning and dissemination of information to the rural communities. Msoffe and Ngulube (2016) add that farmers tend to value information that has a local focus and positive impacts in their farming activities. Thus, it is imperative to understand the information needs and seeking patterns of the target community in order to design an information service that is context specific and relevant to the farmers.

Weather Patterns

Climate condition changes such as change in rainfall amounts, temperature, seasonal patterns and emergency of pests and diseases can cause instability in production and decline in productivity among farmers in the world and Kenya in specific. The effects of these climatic changes requires that farmers access and use of quality climate forecasts and their likely effects on their agricultural production. This information forms the fundamental basis upon which many agricultural decisions relating to when to grow, harvest, apply fertilizer, pesticide, when to stock and destock are made.

Wamalwa, Mburu & Mang'uriu (2016) point out that integration of climate information in agriculture has been undertaken by CARE Kenya, Kenya Meteorological services and ASDSP in all 47 counties in Kenya in the last 5 years. This has been undertaken through an innovative and all-inclusive process referred to as Participatory Scenario Planning. This is where climate forecasts are shared, interpreted and advisories regarding different scenarios and appropriate climate smart practices are generated and disseminated through barazas, radio and normal extension services. However, little is known about the extent of access, knowledge and use of this information by small scale farmers, and management implications in terms of adoption of climate smart practices.

Access and use of weather information has been hampered by a number of challenges such as limitations in terms of information delivery mechanisms, lack of readily available good quality climatic data and climate information to farmers & planners in most developing countries.

Wamalwa, Mburu & Mang'uriu (2016) suggest there is need to strengthen and enhance capacity for collection, downscaling and disseminating agro weather and climate information by meteorological services and extension service providers in a timely manner in order to ensure effective dissemination of agro climate and weather information for enhanced adoption of climate smart practices. They further point out that climate information packaging should be in user friendly formats and up scaling of participatory process involving farmers, local forecasters, meteorological services and extension service providers. This according to them will enhance skill and knowledge in the use of climate and weather information in farming activities.

Pest and Disease Control

Food security and healthcare are the biggest challenges for humanity in the twenty-first century. Utilizing information and communication technologies (ICT) and web-based software systems in agriculture and economy can boost production in the shortest and fastest way. The approach to the development of agriculture apps for farmers is to recognize a clear understanding on agri-business and skill and raise profits for the farming community. The ICT reduces transportation/travel costs and drudgery, transactional and corruption wastes. In particular, in developing countries, ICT can deliver a significant portion of the world's agriculture, improve efficiency levels, and reduce costs and adverse environmental impact of agriculture.

If farmers do not control pests and disease in their farms, they can adversely affect agricultural practices and natural resources, imposing significant economic and environmental costs. They have hence to get information on pest and disease management in order to be able to secure their productions. Multiple applications in crop protection are possible with mobile phones. These are enumerated in this chapter.

Marketing Opportunities for Farm Produce

Market information is an important facilitating function in the agriculture marketing system. It facilitates marketing decisions, regulates the competitive market process and simplifies marketing mechanisms. Market information is a means of increasing the efficiency of marketing system and promoting improved price formation. It is crucial to the farmers to make informed decisions about what to grow, when to harvest, to which market produce should be sent and whether or not to store it. Improved information should enable traders to move produce profitably from a surplus to a deficit market and to make decisions about the viability of carrying out storage where technically possible.

Adopting information and communication technology (ICT) in agricultural extension is crucial in facilitating farmers to access new markets for their produce and acquire information on the current trend in agriculture. Irungu, Mbugua & Muia (2015) argue that it is essential to digitize agricultural production and marketing information into web-based resources. This would enable wider outreach and use since the few available extension officers do not effectively reach the majority of the farmers at different locations. Farmers in Kenya are in desperate need of information on existing and potential markets that would offer good prices for their produce. They also require frequent information on available input and credit markets, especially those that provide accessibility to loan facilities that would enable them procure farm inputs. They need to know who the potential buyers are, where and how to get to them and how to sell their produce at a profit. Information has to be disseminated to the farmers immediately using information technology. Universities and other research institutions in Kenya must play a major role in coming up with new technologies for disseminating market related information to farmers.

Adoption of Technology in Agriculture in Rural Kenya

Transformation in agriculture requires embracing technology in providing access to information through the use of mobile phones and internet. This will influence the growth process of crops to increase the quantity and quality of agricultural production. Information and Communication Technology (ICT) platforms like agricultural extension services provide farmers in rural areas with access to agricultural knowledge such as best practices and market opportunities through the internet and mobile devices. Msoffe & Ngulube (2016) maintain that effective dissemination of agricultural information plays a key role in supporting rural agricultural activities as it may improve farmers' knowledge and increase their agricultural productivity.

The impact of technology on agricultural production highly depend on the level of technology acceptance by farmers and the types of technologies adopted. There are however factors that determine the adoption and use of technology such as ease of its use, social, emotional, attitudinal and cognitive factors. Additionally, technology could be used to grow and enhance agricultural output in rural areas by getting relevant information on best farming inputs. However, lack of basic ICT skills, inadequate political will, insufficient power supply, poor internet connectivity, lack of awareness especially among the rural farmers, absence of agricultural content in local languages and insufficient personnel to handle ICT facilities among other factors may work against the success of agriculture in rural areas. Mobile based technologies have been instrumental in the provision of support to the agricultural sector by provision of digital extension services and data driven decision making within the entire agricultural ecosystem

Agricultural Technologies in Kenya

Digital platforms are playing a vital role in preserving and sharing traditional agricultural knowledge alongside new techniques. This creates a dynamic information exchange where farmers can learn from each other's experiences and make informed decisions tailored to their specific circumstances. These platforms can also connect farmers to agricultural experts and research institutions, fostering knowledge transfer and innovation across the continent.

In Kenya, many small-scale rural farmers lack access to the resources and technical knowledge needed to improve their agricultural productivity. Technological innovations available address opportunities, challenges, and potential solutions to effective agricultural output. Smart agriculture employs modern technology to improve efficiency, sustainability, and excellent farming. This includes use of technological tools that are key in weather forecast remote monitoring, linking farmers with mobile financing services. sensors to track soil health, GPS-based animal care, predictive disease control, and other innovations. In Kenya the following technological innovations have been key to agricultural productivity:

Precision Agriculture

Precision agriculture involves the use of technologies such as GPS, sensors, drones, and data analytics to optimize inputs like water, fertilizers, and pesticides. By precisely targeting inputs based on variations in soil, weather, and crop conditions, farmers can maximize yields while minimizing waste.

Smart Irrigation Systems

Smart irrigation systems, including drip irrigation and sensor-based irrigation, help farmers optimize water use by delivering the right amount of water directly to plant roots based on real-time data on soil moisture levels and weather forecasts.

Additionally, there are specific mobile and internet apps that are instrumental in giving farmers access to a wealth of information on various agricultural practices such as the right crops for their soil conditions and controlling pests. These apps also connect farmers directly to suppliers and consumers, reducing reliance on middlemen and increasing their profits. These include the innovations discussed below:

DigiFarm

This is a technological mobile platform that offers farmers agriculture related information such as best farming practice, agricultural credit services, quality farm input, weather forecasts and market opportunities for farmers. *DigiFarm* helps agribusinesses and small holding farmers share information and transact more easily. It's a technological application provided by Safaricom that offers advice to small holder farmers on where to access quality farm input and market opportunities for the farm produce. It leverages on technology to provide e-extension services through walking with the farmer from the time of planting to harvesting. The technology is aimed to overcome the challenges that small holder farmers face and make farming a profitable venture. *DigiFarm* is a single platform that allows a smallholder farmer access any information that will allow them to have high agricultural production. The e-extension service on *DigiFarm* is a mobile enabled service where the farm indicates the day they plant through an SMS and it triggers a tutorial in terms of the farmer is supposed to be doing from the day they plant until they harvest.

iCow

It is a mobile phone based agricultural information platform for small holder farmers. It is a technology developed to assist animal farmers to overcome challenges of farming.

iCow is an SMS and voice-based mobile phone application for small-scale dairy farmers in Kenya. It's a virtual extension service provider in terms of helping farmers track stages of their cows' growth by giving them valuable tips on cow breeding, animal nutrition, milk production efficiency and gestation. It is an app that aims to reduce cow mortality rates and educate farmers on proper agricultural practices.

Farmers using this technology are required to register for the app on their phones by using the USSD *285# and following the instructions to key in important information about their cows, such as expected calving date and the weight of the cow. The farmers then start receiving timely and tailored messages, advice and information. This app focuses on giving farmers the information and services they need to keep their livestock healthy and increase their milk production.

M-farm

M-Farm, is a transparency tool for Kenyan farmers where they simply SMS the number 20255 (Safaricom Users) to get information pertaining to the retail price of their products, buy their farm inputs directly from manufacturers at favorable prices, and find buyers for their produce. It is an SMS and web-based application focused on improving weaknesses in the value chain. It is a transparency tool for Kenyan farmers to get information pertaining to the retail price of their products, buy their farm inputs directly from manufacturers at favorable prices, and find buyers for their produce. It helps farmers nationwide to access market price data for their crops, ensuring they get a fair price for their product. It is a web and SMS-based application that periodically sends farmers the prices of goods on the market, allowing them to communicate directly with consumers and reducing the power of the middlemen. It gives farmers access to a reliable and guaranteed market that enjoys stable year-round prices while eliminating middlemen and lowering transaction costs

This app is important especially to the rural farmers because eliminates the marketing challenges of rural farmers in Kenya. Rural small hold farmers are always unaware of the market of the various commodities, they produce in excess and are faced with the problem of getting the worth of their produce. Poor information on farm inputs and lack of access to these inputs such as fertilizer, seed, agrochemicals and other equipment are some of the challenges that reduce farmers' productivity in rural areas. The inability of the farmers to have their produce to the markets after harvesting leads them to be exploited by middlemen who offer meager prices for the produce, and even delay payments for the commodities. Offering information to farmers about markets, prices, and other agricultural aspects via mobile(SMS) and web based system enable them to inquire current market prices of different crops from different markets and connect with ready market.

MbeguChoice

Increasing Kenya's food production is the best way to feed our growing population, and to properly and adequately tackle the problem of food insecurity. However, many rural farmers in Kenya may be unaware of new and modern varieties of crop seed that have been bred by local Kenyan breeders for the changing climate conditions and that can produce much higher yields. ***MbeguChoice*** is a mobile application that helps farmers arrive at the best variety of seeds to plant in line with their environmental and climate needs. The farmer may also desire a seed variety with certain attributes suited for their region such as early maturity and drought tolerance, the app is capable of giving this information. The farmer is required to answer questions to ascertain the weather conditions, location of the farmer and drought tolerant varieties, the app then suggests the most suitable variety for the farmer .This technological innovation was developed to bridge the information gap that exists between seed breeders and Kenyan farmers.

MbeguChoice is an online tool that provides information on the wide range of certified crop seed varieties that are available to farmers in Kenya such as the newest releases of seed varieties and those that are best suited to a farmer's region. The app also informs the farmers which seed companies or suppliers should be contacted in case of need so as to make sourcing easy to the farmers.

System of Rice Intensification (SRI)

Rice-growing areas in Kenya can now adopt the System of Rice Intensification (SRI) system to help them save on water, labour costs and seeds while doubling their yields. This low water, labor-intensive, organic method uses younger seedlings singly spaced and typically hand weeded with special tools unlike the traditional flooding method that demanded farmers to transplant only overgrown seedlings, instead of one young seedling hence seed wastage. The old method of rice growing requires a lot of water, making farmers spend a lot of money per planting. SRI involves among its practices, the alternate wetting and drying of paddies, wider spacing and transplanting only one seedling per hill as well as mechanical weeding.

Experts believe that with SRI technology, rice production is set to double since seed wastage would reduce by 90 percent and water costs cut by 40 percent, as the new precise production method will significantly reduce plant population, improve soil conditions and less water will be required. **P8/ Mageso** a farmer in Budalangi, Busia county who has used this technology says '*since adopting this new technique of growing rice, I have no regrets for deciding to undertake rice farming because it has changed my life. I have been able to educate my children without struggling so much as I used before when I entirely relied on bursaries which were not reliable.*

Throughout the country, the shortage of water and land suited for rice production means that extensive expansion of rice growing farmlands is not a likely option. There is therefore need to consider water saving alternatives and any intervention that can increase the productivity of rice and also save on water is a most welcome initiative. The System of Rice Intensification (SRI) therefore offers this opportunity to improve food security through increased rice productivity, increase smallholder farmers' income, water savings and reduce the national rice import bill.

SRI is a set of practices especially developed to improve the productivity of rice grown in paddies. Contrary to the traditional method of continuous flooding of paddy fields, SRI involves intermittent wetting and drying of paddies as well as specific soil and agronomic management practices. It involves some combination of the following changes in rice planting practices; transplant seedlings at a very young age, raise seedlings in un-flooded nurseries, not planted densely and well-supplied with organic matter, transplant seedlings quickly, carefully and shallow, transplant seedlings at wider distance and singly (one seedling only), do not continuously flood the soil, weed control is preferably done with a simple mechanical hand weeding and provide as much organic matter as possible to the soil.

SRI as a technique of rice farming is said to have the following advantages; gives higher yields, require less seed and less water, makes use of what the farmer has (the may not be necessarily required to purchase any extra external inputs) and SRI does not require the purchase of new seeds – since practically all rice varieties give higher yield with SRI.

Impact of Embracing Technology in Agriculture in Rural Areas

The development and adoption of agricultural technology among small hold and rural farmers can play a central role in improving agriculture productivity and sustainability. Technology based applications such as use of sensors, mobile apps, and weather stations to collect real-time information on soil conditions, moisture levels, and weather patterns can empower farmers to make informed decisions on resource utilization. For instance, sensors can determine the specific fertilizer needs for different areas of a field, reducing waste and optimizing crop yields.

Muzari, Gatsi & Muvhunzi (2012) argue that increased technology development and adoption can raise agricultural output, hence improve household food intake. Technological applications give farmers access to a wealth of information on everything from choosing the right crops for their soil conditions to identifying and controlling pests. Additionally, apps connect farmers directly to suppliers and consumers, reducing reliance on middlemen and increasing their profits.

Farmers in rural Kenya are in dire need of information on existing and potential markets that would offer valuable prices for their produce. They need to know who the potential buyers are, where they are, how to get to them and how much they would sell their produce profitably. The ICT initiatives like M-farm have been successful in eliminating the role of intermediaries by acting as marketing channels, which includes transparent pricing. It also provides monetary benefits to farmers compared with the conventional marketing which are costly in terms of transport and identifying a genuine market. These make farmers to have informed decisions of where to sell the crop produce to gain maximum profit. This helps in improving livelihood security in rural areas. Using existing information and communication technologies (ICT) can improve business and networking between farmers, buyers and extension agents and also facilitate access to hidden markets. There is no doubt that enabling access to new technology will play very productive role in ensuring sustainable agricultural production and marketing in rural areas.

Muzari, Gatsi & Muvhunzi (2012) point out that agricultural technology development is an essential strategy for increasing agricultural productivity, achieving food self-sufficiency and alleviating poverty and food insecurity among smallholder farmers in sub-Saharan Africa. They add that this strategy is particularly relevant for the smallholders in the sub-region because they are disadvantaged in many ways, which makes them a priority for development efforts. These are farmers who live and farm in areas where rainfall is low and erratic, and soils tend to be infertile. Information technology can effectively support better seed, fertilizer and pesticide selection to ensure quality and high produce. It also improves land and water management, provides access to weather information, and connects farmers to different markets and sources of credit.

THEORETICAL FRAMEWORK

Diffusion of Innovations theory

Theory of Diffusion of Innovations (DIT), by Everett Rogers (1995), seeks to explain why, how and at what rate new technologies and innovations spread. Diffusion of innovation refers to the process that occurs as people adopt a new idea, product, practice, philosophy, and so on over time, the innovative idea or product becomes diffused amongst the population until a saturation point is achieved. This theory is often regarded as a valuable change model for guiding technological innovation where the innovation itself is modified and presented in ways that meet the needs across all levels of adopters. It also stresses the importance of communication and peer networking within the adoption process.

Rogers (1995) perceives the following as the attributes of an innovation; the type of innovation decision, the nature of communication channel diffusing the innovation at various stages in the innovation decision process, the nature of the social system in which the innovation is diffusing and the extent of change agents' promotion effort in diffusing the innovation. They further argue that the type of innovation –decision is related to the innovation's rate of adoption. Innovations requiring an individual-optional innovation decision are generally adopted more rapidly than when an innovation is adopted by an organization. The more persons involved in making an innovation- decision, the slower the rate of adoption.

Diffusion of innovations is a theory that seeks to explain how, why, and at what rate new ideas and technology spread. It refers to the process that occurs as people adopt a new idea, product, practice, philosophy, and over time, the innovative idea or product becomes diffused amongst the population until a saturation point is achieved.

Simin & Janković (2014) argue that agriculture has always been a specific economic activity which has specific characteristics associated with knowledge, innovation and transfer of new technologies within the knowledge and information system.

Simin & Janković (2014) view diffusion as a special form of communication related to new ideas. It is a specific form of social change, defined as a process by which alteration occurs in the structure and function of a social system while innovation as a social construction created in interaction of awareness and the need for innovation, openness and focus on creating a system of social innovation, creative personalities. They further state that the adoption of innovations in agriculture is positively correlated with the level of education of the adoption unit (farmer), the experience and the property of holdings.

Simin & Janković (2014) observe that the problem of diffusion and implementation of innovations in agriculture should not be considered simplistic, so that one would possibly think that the process of diffusion and implementation of innovation will take place successfully if there are sufficient financial resources, agricultural experts, awareness of adopters, access to innovation etc. This is because in the last century, the experiences of many less developed countries have often proved unsuccessful in modernization of agriculture and rural development although the activities were carefully prepared, generously financed and supported otherwise by the governments of these countries and influential international organizations.

DATA ANALYSIS AND DISCUSSION

This section considers important data to addressing the question of digitizing agricultural production in Kenya. Data presented to support our arguments in this paper were primarily collected from previous publications and interviews from farmers in the western region of Kenya. The interviews were carried out in order to understand how farmers have been able to adopt new technologies in in the production of the agricultural produce and the benefits that are drawn from adopting technology. A total of 13 farmers were interviewed; 8 men and 5 women aged between 25 and 62 years. The participants were drawn from four counties of western Kenya; these are Trans Nzoia, Bungoma, Kakamega and Busia counties.

Table 1. Distribution of participants

County	Number of participants
Trans Nzoia	5
Bungoma	4
Kakamega	2
Busia	2
Total	**13**

The above table shows that most of the participants who took part in the study were from Trans Nzoia county (5) with the least number coming from Kakamega and Busia counties (2).

Table 2. Participants' profiles

S/No	Pseudonym	Gender	Age(Years)	Farming type (Crop/Animal)	Period as farmer (Years)
1	P1/Ken	Male	43	Crop	13
2	P2/Tina	Female	36	Crop	7
3	P3/Melo	Female	47	Crop	17
4	P4/Willy	Male	27	Animal	4
5	P5/Amuke	Male	62	Crop & Animal	21

continued on following page

Table 2. Continued

S/No	Pseudonym	Gender	Age(Years)	Farming type (Crop/Animal)	Period as farmer (Years)
6	P6/Hosea	Male	51	Crop & Animal	15
7	P7/Evano	Male	25	Animal	2
8	P8/ Mageso	Male	45	Crop	12
9	P9/Mary	Female	33	Crop	7
10	P10/Solo	Male	29	Animal	5
11	P11/Lenah	Female	34	Animal	3
12	P12/Rose	Female	31	Crop	6
13	P13/Jonny	Male	41	Crop & Animal	16

The study sought to investigate on how the participants felt about adopting technology in their farming activities. Farmers who were identified through snowballing sampling technique. The farmers had various view in regard to technology in farming:

P6/Hosea one of the farmers from Bungoma says *'Initially I was incurring a lot of losses for I didn't know where to get quality seedlings, pesticide and even fertilizer. Since I was introduced to DigiFarm, my farming has improved. I have benefitted from e-extension services in terms of knowing what to farm in my area, quality inputs to use, where to access inputs on credit. I have so far enrolled five of my neighbours into DigiFarm and they appreciate.*

Information technology empowers farmers to make informed decisions about resource use, planting choices, and harvesting times. One of the farmers from Kitale acknowledges how adoption of technology assisted her to have high productions from her cow.

P2/Tina, a farmer from Kitale says *'iCow has assisted me fetch a lot of money from my cows. If you strictly adhere to the instructions from the app, you will end up with healthier and productive animals. There is a lot I have learnt from this app about diseases, breeding, and new feeding techniques of animal husbandry'*.

Additionally, **P6/Hosea,** a farmer from Bungoma County who has embraced the same technology says: *"iCow has enabled me to get regular* *tips and updates about diseases, breeding, new feeding techniques and best cow-rearing practices'*

Mobile apps are bridging the information gap for farmers by using them for market information, weather forecasts, and even mobile banking for financial transactions. Platforms like M-Farm allow farmers to access market prices, connect with buyers, and exchange knowledge with other farmers through SMS or smartphone apps. **P/7 Evano,** a young farmer in Busia says:

> *'These middlemen often take advantage of our ignorance on the market prices of our products. They purchase our fruits and vegetables at very low prices and re-sell them at high prices, profiting on their sale to the disadvantage us of the producers. Thank you M-Farm, we have realized that the middlemen were exploiting us and we now understand the potentials that our phones can offer. We can now access the market prices and sell directly to consumers at fair value'*

Mobile and web applications for collecting and sending out agricultural market information using simple text messages. These platforms can also be used by individual farmers, agri-businesses and organizations in agricultural value chains to negotiate better prices and choose different markets.

The variable rainfall patterns in Kenya may not promise excellent results, irrigation technologies such as SRI can play a crucial role in ensuring consistent crop production as opposed to the traditional methods previously used. **P8/ Mageso** a farmer from Budalangi points out that *'if more farmers adopt the SRI technology of rice growing, then Bunyala rice irrigation scheme has the potential to become the food basket of the entire western region.'*

P9/Mary another rice farmer says *'I am happy with this new farming method, there are some changes compared with previous years when we had to use traditional ways of planting rice. It was very unproductive unlike now where we have plenty of rice for sale, thanks to SRI.'*

P1/Ken a 43-year-old farmer says: *'Before I knew of* ***DigiFarm****, I depended on the traditional ways of looking for market for my produce and this led to a lot of exploitation. But since the discovery of this app, I have a lot of options for market.'*

This clearly shows that through the use of ICT, farmers can create new opportunities by penetrating even international markets and get contact with new partners and exchange relevant information for their business sustainability.

Information and Communication Technology plays important role in enhancing the impact and performance of agricultural production which could facilitate self-reliance for national growth.

P4/Willy argues that *'Information and Communication Technology (ICT) platforms just like the previously used extension services provide farmers with access to agricultural knowledge on best practices through the internet and mobile devices. Applications such as* ***iCow*** *have enabled high production and ensure food stability.'*

P12/Rose a 23-year-old farmer specializing in in horticulture maintains that *'smart agriculture turns out to be a solution to farmers as it reduces costs, the use of fertiliser and water, which is critical requirements for agricultural productivity.' "Adopting technology has contributed reduced costs of production and increased income levels. We use climate-smart irrigation solutions to help combat the effects of climate change which increasingly affect our crops and bio-pesticides to fight emerging pests and diseases. We are happy since we are able to meet our daily family obligations and equally have managed to start other income-generating schemes in addition to agriculture"* added **P12/Rose**

ICT offer a great opportunity in transferring knowledge to farmers, advising and educating farmers in their decision making, enabling farmers to clarify their own goals and possibilities, and stimulating desirable agricultural developments. This ensures that farmers improve crop yields and livestock productivity, contributing to food security and income generation.

ICTs is an instrument of empowerment for smallholder farmers, allowing them to follow the cycle of production from purchasing seeds to the sale of their products. **P5/Amuke**, one of the oldest farmers interviewed comments thus: *The biggest problem that technology has solved for farmers is providing adequate market security for both farmers and vendors so that farmers can access adequate market for their produce. For a long time, my produce had gone to waste due to lack of market.*

The researcher's interaction with the farmers generally showed that technology has been able to uplift the farmers 'status in rural areas. To sample, a few narrations from the farmers were as follows: **P8/ Mageso** says that .... *through technology, small-scale farmers in rural areas learn to use instruments like their cell phones, computers, and video-cameras to access information and find alternative ways to sell their products,"*. These sentiments are echoed by **P11/Lenah** who observes that ......*farmers in rural areas far away from urban centres can now access ICT-based services which offers them the possibility to access information that is often monopolized by intermediaries as such, farmers are able to use sensors to track soil health, choose the best seeds and fertilizer for the soil type, and monitor plant growth.* **P13/Jonny** a 41 years old animal farmer maintains that ...... *these innovations have improved efficiency in the supply chain and curtailed post-harvest losses.* Mobile technology can help farmers access the services whenever they need, **P10/Solo** says....... *technology through precision agriculture facilitates management of farm inputs such as fertilizer, seeds, and water, based on real-time information.* And just like the rest **P7/Evano** maintains that ......*Kenya is tapping into innovation and technology from multiple angles to build a model farming system. We have innovations that help generate agriculture-related advisories. More farmers are using technology to get timely information that can inform quick responses.*

Technology connects rural farmers to relevant information, quality inputs and credible agriculture services accessible to them in the local community. P3/Melo confirms 'Technology has really been of value to me, *my farm has become land mark for agricultural change in this community and I am now a respected farmer. Every day, other farmers come to my farm to learn one or two things from my farm.*

CONCLUSION

Adoption of technologies in agricultural production is a surety to sustainable agricultural production, however, it is essential to ensure that these technologies are accessible and affordable to a majority of farmers, both small scale and large scale. It is therefore important to train these farmers on the available technological innovation that they can be able to access and use in order to enhance their agricultural production.

Agriculture is entering a new era where digital tools shape how farmers grow crops, rear animals and reach consumers after harvesting virtually.

Innovation in agriculture is leading an evolution in agricultural practices, thereby reducing losses and increasing efficiency. Technology plays a crucial role in the agriculture industry by helping farmers overcome various challenges they face in their day-to-day operations. It provides farmers with access to agricultural knowledge, best practices, and market opportunities through the internet and mobile devices. As a result, farmers have realized increased production and safe growing conditions. Due technological innovation, farmers have realized increased crop production and a reduction in water, fertilizer, and pesticide application.

Adoption of technology has ensured development of cost-effective solutions for agricultural production, post-harvest and storage processing, including solar pumping, cooling, chilling and drying. These technologies result in saved costs and increased yields for local farmers.

Digital platforms such as iCow are linking farmers with timely and accurate developments in science and research. Timely and accurate agriculture-oriented advisories are boosting production and enabling farmers to sustain the livestock sub-sector which is a major backbone to Kenya's economy.

The changing weather patterns are making it difficult for farmers in the country to plan using the traditional knowledge. Technological innovations such as irrigation technology in agriculture is giving farmers new ways of rain-sensing technology, and soil moisture sensors are specifically designed to track certain irrigation intervals even when it is raining and water content in the soil to determine irrigation needs. Farmers are urged to fully adopt irrigation farming as opposed to rain-fed farming that is prone to poor harvest. Farmers must re-consider adopting irrigation farming to utilize available land and help address the issue of food shortage that has been witnessed in Kenya for a long time.

As can be witnessed from the foregoing discussion clearly indicates that adoption of innovations and new technologies in agriculture can have a central contribution to the achievement of food security. Various innovative ICT applications by both government and private initiatives have been developed for better communication and rapidly changing demand of the consumers in rural areas. These applications are playing a vital role in preserving and sharing traditional agricultural knowledge alongside new techniques. This creates a dynamic information exchange where farmers can learn from each other's experiences and make informed decisions tailored to their specific circumstances. As ICT access continues to increase among farming communities and information services continue to adapt and flourish, the scope exists for a much better rural productivity impact in the future.

REFERENCES

Abebaw, A. A. (2020). Challenges and Opportunities of Information and Communication Technologies for Dissemination of Agricultural Information in Ethiopia. *In International Journal of Agricultural Extension*: https://www.esciencepress.net/IJAE

Chelule, R. C., Munyua, C. N., & Kibett, J. (2022). Perceptions of Agricultural Extension Staff towards Financial Support Before and After Devolution in Kericho County, Kenya. In *Asian Journal of Agricultural Extension, Economics & Sociology.40(9): 128-135, 2022; Article no. AJAEES.87817 ISSN: 2320-702*

Hamid, E. B., Tarek, B. H., Francesco, B., Sinisa, B., & Roberto, C. (2021). Acceptance and Adoption of Technologies in Agriculture. *In AGROFOR International Journal,* Vol. 6, Issue No. 1, 2021

Irungu, K. R. G., Mbugua, D., & Muia, J. (2015). Information and Communication Technologies (ICTs) Attract Youth into Profitable Agriculture in Kenya. *East African Agricultural and Forestry Journal, 81*(1), 24–33. DOI: 10.1080/00128325.2015.1040645

Kukreja, A., & Chakrabarti, B. (2013).Agricultural Knowledge Management and Dissemination: Initiatives by Information and Communication Technology. *In Journal of Global Communication.* Vol. 6, No. 1, Jan. - June. 2013: 51-58. DOI: DOI: 10.5958/j.0976-2442.6.1.006

Msoffe, G., & Ngulube, P. (2016).Agricultural Information Dissemination in Rural Areas of Developing Countries: A Proposed Model for Tanzania. *In African Journal of Library, Archives and Information Science*: https://www.researchgate.net/publication/312070206

Muzari, G., Gatsi, W., & Muvhunzi, S. (2012). The Impacts of Technology Adoption on Smallholder Agricultural Productivity in Sub-Saharan Africa: A Review. *Journal of Sustainable Development, 5*(8), 2012.

Oyindeinbofa, N. (2018). Dissemination of Agricultural Information to Farmers using ICT. In *International Journal of Computer Applications* (0975 – 8887) Volume 179 – No.7, December 2017. Kobe Institute of Computing, Graduate School.

Rogers, E. M. (1995). *Diffusion of Innovations* (4th ed.). The Free Press.

Servaes, J., & Lie, R. (2020). Key Concepts, Disciplines, and Fields in Communication for Development and Social Change. In Servaes, J. (Ed.), *Handbook of Communication for Development and Social Change*. Springer Nature Singapore Pte Ltd., DOI: 10.1007/978-981-15-2014-3_113

Simin, M. T., & Janković, D. (2014). Applicability of Diffusion of Innovation Theory in Organic Agriculture. In *Economics of Agriculture* 2/2014 UDC: 631.147:001.895

Wamalwa, I. W., Mburu, B. K., & Mang'uriu, D. G. (2016). Agro Climate and Weather Information Dissemination and Its Influence on Adoption of Climate Smart Practices among Small Scale Farmers of Kisii County, Kenya. In *Journal of Biology, Agriculture and Healthcare*. ISSN 2224-3208 (Paper) ISSN 2225-093X (Online), Vol.6, No.10, 2016

Chapter 8
James Webb Space Telescope and Discovery Anxiety:
Role of the Sources of Information

Adebowale Jeremy Adetayo
https://orcid.org/0000-0001-7869-5613
Adeleke University, Nigeria

ABSTRACT

The present chapter aimed to investigate the impact of information sources on anxiety related to potential discoveries by the James Webb Space Telescope (JWST). A sample of 1009 participants was surveyed, and descriptive and inferential statistical analysis was performed. Results indicated that 24.6% of the participants reported anxiety concerning JWST's discoveries, while 41.3% believed in the existence of extraterrestrial intelligence. The chapter revealed a significant correlation between information sources and anxiety towards JWST's discoveries (R = 0.268, R2 = 0.72, Adj R2 = 0.059, P < 0.05). It was discovered that library and online newspaper sources were the major contributors to such anxiety. The findings highlight the significance of information sources in shaping individuals' reactions to potential extraterrestrial discoveries. Therefore,

DOI: 10.4018/978-1-6684-9356-4.ch008

the chapter recommends libraries and online newspapers to educate the public about JWST's findings, alleviating anxiety and promoting accurate information.

INTRODUCTION

The James Webb Space Telescope (JWST) is a huge, space-based astronomical observatory that examines the time when the universe's initial stars and galaxies originated and reionized. It was created by NASA, a prime contractor team led by Northrop Grumman Space Technology, the European Space Agency, and the Canadian Space Agency. The JWST was designed to be a follow-up mission to the extremely successful Hubble Space Telescope, allowing astronomers to glimpse the initial generations of stars. JWST, which has a massive 18-segment, 6.5-meter main mirror and an array of new infrared-sensing cameras and spectrometers, will allow us to observe younger objects in space than Hubble presently can and will aid in the analysis of the minute specks of light that Hubble cannot detect. The JWST is poised to revolutionize our understanding of the cosmos through its comprehensive examination of astronomical phenomena. The JWST has been designed to address four primary themes, each focused on a specific aspect of the universe: First Light and Reionization, Galaxies' Formation, Birth of Stars and Protoplanetary Systems, and Planetary Systems and the Origins of Life.

First Light and Reionization aim to uncover the first galaxies to form in the universe and trace the history of ionization that led to the current state of the cosmos. The Galaxies' Formation theme will shed light on the development of galaxies and dark matter over time. The Birth of Stars and Protoplanetary Systems theme focuses on the earliest stages of star formation and the development of protoplanetary systems. Finally, the Planetary Systems and the Origins of Life theme will examine the physical and chemical properties of both our solar system and extra-solar systems to gain insight into the origins of life. Each of these themes represents a significant area of astronomical inquiry and the JWST is poised to make groundbreaking contributions in each of these areas.

In addition to its capability to explore the far reaches of the universe, the JWST also has the potential to make significant contributions to our understanding of the Milky Way and its neighboring galaxies. By studying ancient stars and star remnants that date back to the early stages of the Milky Way's development, astronomers can gain a deeper understanding of the history and evolution of the galaxy.

The JWST's advanced infrared sensors are particularly well suited to this task. The ability to penetrate the dust and gas that often shrouds stellar nurseries allows scientists to observe and analyze the formation and development of stars in detail. This includes the mechanisms that control star mass and composition, as well as the formation of heavy elements (Sabelhaus & Decker, 2004). These insights are essential for understanding not just the Milky Way, but also the universe as a whole. The JWST is expected to make numerous contributions to the field of astronomy, with its ability to probe the early stages of star and galaxy formation being a key area of focus. By exploring the furthest reaches of the universe and the history of our galaxy, the JWST will greatly enhance our understanding of the cosmos.

The JWST, launched in December 2021, has already established itself as one of the most advanced space telescopes in existence. Its capabilities surpass even those of the Hubble Space Telescope, which has been a cornerstone of astronomical research for decades. The JWST's ability to directly observe exoplanets, or planets orbiting stars other than our own Sun, holds great promise for astronomical discovery. In particular, the potential for the JWST to detect signs of extraterrestrial life has sparked a renewed interest in astrobiology and the societal implications of such a discovery.

This new avenue of research has been made possible through the support of the NASA Astrobiology Program and the John Templeton Foundation. The funding received from these organizations has enabled astronomers to explore the significance of finding life beyond our planet and to consider the impact such a discovery would have on our understanding of the universe (, 2016). It is a truly exciting time for the astronomical community as the JWST begins to survey the cosmos and gather data that has the potential to reshape our understanding of the universe.

However, the abundance of information sources and the rapid spread of news through social media has made it increasingly challenging to separate fact from fiction (Adetayo, 2021; Adetayo et al., 2021). The issue of intelligent life in space is no exception, and the uncertain credibility of various sources of information may affect people's reactions to the discovery of extraterrestrial life. Social media, for example, is a popular and fast-paced platform for news, but its information is not always vetted. Religious leaders have been a trusted source of information for many individuals, and traditional media such as television, radio, and newspapers are also important sources of daily knowledge.

Libraries, too, play a significant role in providing information on topics such as the discovery of extraterrestrial life. As a trusted and objective source of information, libraries offer users a safe space to access and explore materials without the influence of biased interpretations. Librarians are trained to avoid censorship and to provide materials regardless of personal beliefs (Adetayo, 2022). This is why libraries participated in the JWST launch celebration (Jordan-Heintz, 2021).

In light of these considerations, this chapter aims to evaluate the impact of different information sources on anxiety related to the JWST's potential discovery of extraterrestrial life. By exploring the influence of various sources of information, the chapter aims to shed light on the complex relationship between the discovery of alien life and the public's reaction to it.

Objectives of the Chapter

The overarching objective of this investigation is to assess the impact of information sources on anxiety related to potential discoveries by the James Webb Space Telescope. To achieve this objective, the chapter will undertake a series of measures to:

1. Evaluate the sources of information utilized by students in acquiring knowledge about the JWST,
2. Determine the prevalence of anxiety among students regarding the potential discoveries of the JWST,

3. Assess the students' belief in the existence of extraterrestrial intelligence, and
4. Evaluate the impact of the information sources on the students' anxiety concerning the discoveries of the JWST

LITERATURE REVIEW

The James Webb Space Telescope

The James Webb Space Telescope JWST is a next-generation space observatory designed to study the universe in infrared light. It is designed to study a wide range of astronomical objects and phenomena. The main components of the JWST include a main mirror composed of 18 different pieces that unfold and adapt to form after launch. The mirrors are composed of beryllium, which is extremely light. Webb's most notable feature is a five-layer sunshield the size of a tennis court that attenuates heat from the Sun more than a million times. The detectors in the telescope's four instruments, cameras and spectrometers, can record exceedingly low signals. One device (NIRSpec) features programmable micro-shutters that allow it to see up to 100 objects at the same time. Webb also carries a cryocooler for chilling another instrument's mid-infrared detectors (MIRI) to a very low 7 kelvins (minus 447 Fahrenheit) so they can operate (Garner, 2015).

The JWST has several key features and capabilities that make it a unique and powerful astronomical tool. For example, the JWST's large segmented mirror and sunshield allow it to study the universe in infrared light, which is critical for understanding the formation and evolution of galaxies, stars, and planetary systems. Additionally, the JWST has the ability to observe the universe at high spatial and spectral resolution, enabling scientists to obtain detailed information about astronomical objects and phenomena. The JWST's scientific instruments also have the capability to perform observations simultaneously in multiple wavelength bands, which provides a more complete picture of the objects being studied (Royal Museums Greenwich, 2022).

The JWST is a highly anticipated and much-studied astronomical tool. Previous research related to the JWST has focused on a variety of topics, including its design and development, scientific capabilities, and potential science targets (Sabelhaus & Decker, 2004; Wright et al., 2008).

Information Sources

Information sources play a critical role in shaping public perception and understanding of scientific concepts and technological developments. In the context of the JWST, information sources can have a significant impact on public perception and understanding of the potential discoveries and scientific achievements of the JWST.

Information sources are a source of information for someone, i.e. anything that can inform or educate someone about anything (Lisbdnetwork, 2018). These sources can include traditional media outlets such as television and newspapers, as well as digital media such as websites, social media, and online forums. Information sources can also include libraries and religious bodies.

Information sources play a critical role in shaping public perception (Tsoy et al., 2021) and understanding of scientific concepts and technological developments (Lee, 2020; Rosenthal, 2022), including the JWST. The type of information source and the way the information is presented can greatly impact public perception and understanding of the JWST and its potential discoveries. For example, information presented in libraries will likely have a different impact on public perception than information presented in a social media outlet. Additionally, the type of information source can impact the level of detail and accuracy of the information that is communicated to the public.

Anxiety Related to Potential Discoveries

Anxiety is a common emotional response to new and uncertain situations, and it can also be experienced in response to scientific developments and technological advancements. In the context of the JWST, the potential discoveries and scientific achievements of the telescope may lead to feelings of anxiety for some members of the public.

Anxiety is defined as an anticipatory response to uncertain, future threats (Glasgow et al., 2022). Anxiety can be experienced in response to a wide range of situations and events, including new and uncertain situations, such as scientific advancements and technological developments. Anxiety related to scientific and technological advancements is a well-known phenomenon, and it is often referred to as "technological anxiety" (Troisi et al., 2022). This type of anxiety can be experienced in response to new and uncertain scientific developments and technological advancements, such as the potential discoveries of the JWST. Technological anxiety can be driven by a range of factors, including lack of understanding, fear of the unknown, and concerns about the potential consequences of scientific and technological advancements.

Previous research has explored the relationship between anxiety and technological advancements, including scientific developments. For example, studies have investigated the impact of public perceptions of and responses to a wide range of new energy technologies (Boudet, 2019). Other studies have examined the impact of anxiety related to technological advancements on individual well-being (O'Driscoll et al., 2010). These studies provide valuable insights into the ways in which anxiety related to technological advancements can impact individual and societal well-being and the strategies that can be used to mitigate this anxiety.

Information Sources, Anxiety, and JWST

The impact of information sources on anxiety related to potential discoveries by the JWST is a complex and multi-faceted issue. The type of information source, the way the information is presented, and accuracy of the information may all play a role in shaping anxiety related to the JWST and its potential discoveries. Information sources may play a critical role in shaping public perception and understanding of the JWST and its potential discoveries, and they may also impact anxiety related to the JWST. For example, if the information about the JWST and its potential discoveries is presented in a sensationalized manner, it may contribute to feelings of anxiety for some members of the public. On the other hand, if the information about the JWST is presented clearly and educationally, it may help to reduce anxiety and promote understanding. The type of information source, the way the information is presented, and the level of detail and accuracy of the information can all play a role in shaping anxiety related to the JWST and its potential discoveries.

METHODS

Settings

{Anonymized} is a private and faith-based learning institution situated on 520 acres of land in {Anonymized}, Nigeria, which was established by the {Anonymized} in 2011. The Foundation is a philanthropic, non-profit organisation devoted to providing impoverished students with access to high-quality higher education based on the {Anonymized} education philosophy, which emphasises healthy living. The university has six faculties, which include Faculty of Science, Faculty of Basic and Medical Science, Faculty of Business and Social Science, Faculty of Art, Faculty of Law, Faculty of Engineering.

Research Design

The present chapter employed a descriptive survey research design in order to investigate the relationship between information sources and anxiety about the potential discoveries of the JWST. The survey research method was deemed an appropriate choice for the study as it provided a comprehensive and reliable depiction of the role of information sources in shaping individuals' attitudes towards the JWST's discoveries.

Population and Sampling Size

The study population comprised of 3457 undergraduate students at {Anonymized}. A simple random sampling technique was used to select 30% of students from all departments, resulting in a sample size of 1037.

Research Instrument

The chapter utilized a structured questionnaire as the research instrument. Questionnaires are a suitable tool for conducting surveys as they allow for efficient data collection from a large number of individuals. The questionnaire underwent a review process by an expert in the field of information management to establish its face validity. Based on the expert's suggestions, some sections of the questionnaire were revised and clarified to improve the accuracy of the responses. The reliability of the instrument was established through a test-re-test method and was reported to have an internal consistency of $\alpha = 0.83$.

Method of Data Collection

The validated questionnaires were distributed among 1037 undergraduate students from the selected departments at {Anonymized}. Participants were informed that their responses would be kept confidential and used only for academic research purposes, and that no third party would have access to their information. A total of 1009 students completed the questionnaire, representing a response rate of 97%.

Method of Data Analysis

The collected data were analyzed using both descriptive and inferential statistics. Descriptive statistics such as frequency counts, percentages, mean, and standard deviation were utilized to answer the research questions. Inferential statistics, specifically multiple regression, were used to evaluate the single hypothesis proposed in the chapter.

FINDINGS

This section presents the findings of the chapter using descriptive and inferential statistics. The details are illustrated in Table 1-5.

Table 1. Socio-demographic characteristics of the respondents

Socio-Demographic Characteristics	Categories	Frequency	Percent
Age Range	11-15	37	3.7
	16-20	754	74.7
	21-25	136	13.5
	26-30	27	2.7
	31 & above	55	5.5
Gender	Female	602	59.7
	Male	407	40.3
	Total	1009	100.0

The findings of this chapter reveal that the majority of the respondents were young and female, with 59.7% of participants falling within the age range of 16-20 years old. This indicates that the chapter sample was primarily composed of youthful and feminine individuals.

Table 2. Sources of Information about James Webb Space Telescope

S/N	Sources	Yes	No
1.	Facebook	213(21.1%)	796(78.9%)
2.	Twitter	121(12.0%)	888(88.0%)
3.	Instagram	171(16.9%)	838(83.1%)
4.	YouTube	198(19.6%)	811(80.4%)
5.	Television	205(20.3%)	804(79.7%)
6.	Radio	121(12.0%)	888(88.0%)
7.	Newspapers	169(16.7%)	840(83.3%)
8.	Church/Mosque	80(7.9%)	929(92.1%)
9.	Blogs	204(20.2%)	805(79.8%)
10.	Google	353(35.0%)	656(65.0%)
11.	Family	120(11.9%)	889(88.1%)
12.	Friends	219(21.7%)	790(78.3%)
13.	Religious Leaders	87(8.6%)	922(91.4%)
14.	Library	264(26.2%)	745(73.8%)

In terms of information sourcing about the JWST, the results presented in Table 2 demonstrate that a considerable proportion of respondents did not source information from any specific platform. However, 35% of participants reported using Google as their primary source of information, while 26.2% of respondents used library facilities to gather information about the JWST. On the other hand, religious leaders and religious institutions such as churches and mosques were the least utilized sources, accounting for only 8.6% and 7.9% of participants, respectively. These findings underscore the importance of exploring the different platforms used by individuals to source information about the JWST, as they may play a significant role in shaping their perceptions and reactions to the telescope's discoveries.

Table 3. Anxiety of what the James Webb Telescope will discover

Prevalence of Anxiety	Frequency	Percent
No	761	75.4
Yes	248	24.6
Total	1009	100.0

The findings from Table 3 indicate that a majority of respondents (75.4%) are not anxious about what the JWST may discover in the future. Conversely, 24.6% of the participants expressed anxiety regarding the potential discoveries of JWST. These results suggest that the majority of the respondents have a positive outlook towards the new knowledge and discoveries that JWST may bring to humanity.

Table 4. Belief in the existence of intelligent aliens in space

	Frequency	Percent
Not Sure	385	38.2
No	207	20.5
Yes	417	41.3
Total	1009	100.0

In Table 4, the results showed that 41.3% of the participants believe in the existence of intelligent life in space. This result implies that a significant number of respondents already have a preconceived idea that extraterrestrial life may exist. On the other hand, 38.2% of the participants were undecided about this issue, suggesting that they may require more concrete evidence before making a firm belief.

Overall, these findings suggest that while a portion of the respondents has strong beliefs in the existence of extraterrestrial life, a significant number of participants may require further evidence and information to solidify their stance. It highlights the ongoing debate and differing perspectives within the scientific community regarding the presence of extraterrestrial life and the potential discoveries that JWST may uncover.

Table 5. Information sources and fear of what the James Webb Telescope will discover

Model Summary						
Model	**R**	**R Square**	**Adjusted R Square**		**Std. Error of the Estimate**	
1	.268[a]	.072	.059		.418	
ANOVAa						
Model		**Sum of Squares**	**df**	**Mean Square**	**F**	**Sig.**
1	Regression	13.410	14	.958	5.483	.000[b]
	Residual	173.635	994	.175		
	Total	187.045	1008			
Coefficientsa						
Model		**Unstandardized Coefficients**		**Standardized Coefficients**	**t**	**Sig.**
		B	**Std. Error**	**Beta**		

continued on following page

Table 5. Continued

1	(Constant)	.856	.070		12.224	.000
	Facebook	.064	.044	.060	1.448	.148
	twitter	-.085	.050	-.064	-1.680	.093
	Instagram	.026	.047	.022	.543	.587
	Youtube	.030	.045	.027	.663	.508
	TV	-.008	.044	-.008	-.190	.849
	Radio	.079	.054	.059	1.453	.146
	Online Newspaper	.139	.049	.121	2.866	.004
	Church/Mosque	.082	.061	.051	1.328	.184
	Blog	-.003	.043	-.003	-.079	.937
	Google	-.015	.037	-.017	-.402	.688
	Family	-.045	.053	-.034	-.848	.396
	Friends	.034	.042	.032	.809	.419
	Religious Leaders	-.077	.063	-.051	-1.238	.216
	Library	.105	.038	.107	2.755	.006

The findings from the regression analysis indicate a moderate, positive relationship between information sources and anxiety towards JWST's discoveries, with an R-value of 0.268, representing 26.8% strength. The ANOVA analysis produced an F-value of 5.483 and a p-value of 0.000, which is statistically significant at the 0.05 level. This suggests that the information sources have a significant impact on the anxiety levels among the chapter participants. The R2 value of 0.072 and the adjusted R2 value of 0.059 indicate that information sources contribute 7.2% and 5.9%, respectively, to the anxiety levels towards JWST's discoveries. It is important to note that among the information sources, library and online newspapers had the greatest impact, as indicated by their significant coefficients in Table 5.

DISCUSSIONS

The JWST, with a cost of $10 billion, has been making headlines with its breathtaking imagery, even before it reaches its full operational capacity. On March 16th, 2022, the JWST team released a spectacular photo capturing a blazing star emitting streams of light, surrounded by an abundance of distant galaxies. This image has captivated the scientific community, as it demonstrates the JWST's ability to penetrate previously hidden regions of the cosmos. The thousands of speckles and spots visible in the photo are each distant galaxies, previously beyond our reach (Pultarova, 2022). The release of this photo has only increased excitement for the JWST's future potential and the groundbreaking discoveries that it will enable.

As awareness of the JWST continues to increase through various media sources, more people are becoming interested in this groundbreaking project. However, this chapter found that a majority of students are not actively seeking information about the JWST. This could be attributed to the COVID-19 pandemic and the related media coverage, as well as the conflict in Ukraine, both of which have captured the world's attention (Apuke & Omar, 2021). Despite this, the chapter found that Google and libraries were the most frequently used sources of information about the JWST among the students surveyed. Although both sources were utilized by less than half of the participants, the trustworthiness of libraries may have influenced their usage (PEW, 2017). The dominance of Google as an information source is unsurprising, given its status as the world's most popular search engine. The chapter highlights the need for continued efforts to increase public awareness and understanding of this exciting project and its potential impact on our understanding of the universe.

The majority of students surveyed in this chapter appear to have little anxiety regarding potential discoveries from the JWST. This could be due to the perception that any new findings from the JWST would have a limited impact on their daily lives, considering that Earth's existence has remained relatively unchanged despite prior discoveries in outer space. However, the chapter also found that nearly half of the respondents (41.3%) believe in the existence of intelligent life in space, despite the lack of secular evidence supporting this claim. On the other hand, 38.2% expressed skepticism about the possibility of intelligent aliens, highlighting the divided opinions on this topic. As the JWST continues to progress towards full operation, it has the potential to shed light on these debates by providing new and compelling evidence regarding the universe and its potential inhabitants. Thus, it is possible that the public's views and beliefs on this topic may shift as the JWST's capabilities continue to unfold.

The current chapter highlights the crucial role that information sources play in shaping people's anxieties about potential discoveries from the JWST. The findings align with previous research by McGee et al. (2021), which demonstrated that different information sources can have varying effects on individual anxieties. In this chapter, online newspapers and libraries were identified as key sources that influence people's anxieties regarding what JWST may discover. This highlights the level of trust that people place in libraries, as they are known to have comprehensive collections of information and reliable sources. However, skepticism about extraterrestrial life in space has long been a dominant theme in library literature related to space research and discoveries. If JWST were to provide evidence that contradicts these skeptical views, it would represent a significant departure from the information traditionally found in libraries. This could explain why individuals who primarily seek information on JWST in libraries may be more concerned about the possibilities of what JWST may reveal, particularly in light of speculation surrounding the existence of alien life in space.

The chapter had some limitations that may have influenced the results. Firstly, the chapter was limited to undergraduate students at {Anonymized}, so the results may not be generalizable to the larger population. Future research could explore other variables that influence anxiety towards what JWST may discover, such as cultural and socioeconomic factors. Additionally, future research could also be done on a larger sample size and in other countries to determine the generalizability of the results.

CONCLUSION

In conclusion, the research findings suggest that information sources play a crucial role in shaping people's reactions and anxieties towards potential discoveries from the JWST. Most students do not actively seek information about JWST, but those who do turn to Google and libraries as their primary sources.

While a majority of individuals are unconcerned about what JWST may uncover, there is a minority who have skeptical views about the existence of intelligent alien life in space. As such, the chapter recommends that libraries and online newspapers continue to provide up-to-date information and educate the public on developments in alien life research. This will help to dispel any anxieties that people may have and keep them informed on the latest advancements in the field. Additionally, the chapter highlights the importance of staying vigilant about the role of information sources in shaping public perception and reactions to major discoveries.

Implications of the Chapter

The chapter has implications for search engines and libraries. The chapter found that Google is the most popular search engine among students seeking information about JWST. As a result, search engines have a responsibility to provide relevant and accurate information about JWST and related subjects. This will ensure that people who turn to search engines as a source of information are not misinformed or misled.: This means that search engines need to be vigilant about algorithmic bias when ranking and displaying information about JWST and related subjects. Search engines should strive to present a diverse range of perspectives and opinions about JWST discoveries, to ensure that people have access to a well-rounded view of the subject. This will help to mitigate any biases that may arise from the ranking and presentation of information in search results.

Also, libraries should be aware of the significant role they play in shaping people's perceptions about what JWST may discover. The high level of trust that patrons have in libraries means that libraries have a unique opportunity to influence their patrons' attitudes towards the possibilities of what JWST may uncover. Moreover, the chapter suggests that libraries should stay up to date with developments in space research and discovery, specifically the JWST project. This will help libraries to effectively educate their patrons on the latest research findings, as well as dispel any myths or misconceptions about the possibility of extraterrestrial life in space. This will help reduce any anxiety that patrons may have and enable them to form their own informed opinions on the matter.

REFERENCES

Adetayo, A. J. (2021). Fake News and Social Media Censorship. In Blankenship, R. J. (Ed.), *Deep Fakes, Fake News, and Misinformation in Online Teaching and Learning Technologies*. IGI Global., DOI: 10.4018/978-1-7998-6474-5.ch004

Adetayo, A. J. (2022). Phronesis and library selection: Dilemmas in Nigerian faith-based universities. *College & Research Libraries News*, *83*(3), 114–117. DOI: 10.5860/crln.83.3.114

Adetayo, A. J., Sanni, B. A., & Aborisade, M. O. (2021). COVID-19 Vaccine Knowledge, Attitude, and Acceptance among Students in Selected Universities in Nigeria. *Dr. Sulaiman Al Habib Medical Journal*, *3*(4). Advance online publication. DOI: 10.2991/DSAHMJ.K.211014.001

Apuke, O. D., & Omar, B. (2021). Television News Coverage of COVID-19 Pandemic in Nigeria: Missed Opportunities to Promote Health Due to Ownership and Politics. *SAGE Open*, *11*(3). Advance online publication. DOI: 10.1177/21582440211032675

Boudet, H. S. (2019). Public perceptions of and responses to new energy technologies. *Nature Energy*, *4*(6), 446–455. DOI: 10.1038/s41560-019-0399-x

Garner, R. (2015, November 20). *James Webb Telescope Overview*. NASA. https://www.nasa.gov/mission_pages/webb/about/index.html

Glasgow, S., Imbriano, G., Jin, J., Zhang, X., & Mohanty, A. (2022). Threat and uncertainty in the face of perceptual decision-making in anxiety. *Journal of Psychopathology and Clinical Science*, *131*(3), 265–277. DOI: 10.1037/ABN0000729 PMID: 35357845

Jordan-Heintz, S. (2021, November 9). *Ericson Public Library to host Webb telescope event Tuesday*. Ames Tribune. https://www.amestrib.com/story/news/2021/11/09/ericson-public-library-host-webb-telescope-event-tuesday/6355507001/

Lee, P. C. (2020). Technological innovation in libraries. *Library Hi Tech*, *39*(2), 574–601. DOI: 10.1108/LHT-07-2020-0163/FULL/XML

Lisbdnetwork. (2018, November 16). *Sources of Information*. https://www.lisedunetwork.com/sources-of-information/

McGee, M. A., Maitland, C., & Njoroge, D. (2021, August 23). Information Sources and Anxiety among Refugees in Kenya during COVID-19. *1st Virtual Conference on Implications of Information and Digital Technologies for Development,* https://doi.org/DOI: 10.48550/arxiv.2108.09935

O'Driscoll, M. P., Brough, P., Timms, C., & Sawang, S. (2010). *Engagement with information and communication technology and psychological well-being* (Vol. 8). Research in Occupational Stress and Well Being, DOI: 10.1108/S1479-3555(2010)0000008010/FULL/XML

PEW. (2017, September 14). *Libraries are the Most Highly Trusted Resource for Information on Recent Pew Survey – Library Research Service*. PEW. https://www.lrs.org/2017/09/14/libraries-are-most-highly-trusted-resource-for-information-on-recent-pew-survey/

Pultarova, T. (2022, March 17). *James Webb Space Telescope's 1st view of the cosmos has scientists thrilled for more*. Space. https://www.space.com/james-webb-space-telescope-better-than-expected-image

Rosenthal, S. (2022). Information sources, perceived personal experience, and climate change beliefs. *Journal of Environmental Psychology, 81*, 101796. DOI: 10.1016/J.JENVP.2022.101796

Royal Museums Greenwich. (2022). *What can the James Webb Space Telescope do?* https://www.rmg.co.uk/stories/topics/james-webb-space-telescope-vs-hubble-space-telescope

Sabelhaus, P. A., & Decker, J. E. (2004). An overview of the James Webb Space Telescope (JWST) project. *Optical, Infrared, and Millimeter Space Telescopes, 5487*(12), 550–563. DOI: 10.1117/12.549895

Troisi, O., Fenza, G., Grimaldi, M., & Loia, F. (2022). Covid-19 sentiments in smart cities: The role of technology anxiety before and during the pandemic. *Computers in Human Behavior, 126*, 106986. DOI: 10.1016/J.CHB.2021.106986 PMID: 34511715

Tsoy, D., Tirasawasdichai, T., & Kurpayanidi, K. I. (2021). Role of Social Media in Shaping Public Risk Perception during COVID-19 Pandemic: A Theoretical Review. *International Journal of Management Science and Business Administration*, *7*(2), 35–41. DOI: 10.18775/IJMSBA.1849-5664-5419.2014.72.1005

Wright, G. S., Reike, G., Barella, P., Boeker, T., Colina, L., van Dishoeck, E., Driggers, P., Goodson, G., Greene, T., Heske, A., Henning, T., Lagage, P.-O., Meixner, M., Norgaard-Nielsen, H., Olofsson, G., Ray, T., Ressler, M., Thatcher, J., Waelkens, C., & Zehnder, A. (2008). Design and development of MIRI, the mid-IR instrument for JWST. *Space Telescopes and Instrumentation 2008. Optical, Infrared, and Millimeter*, *7010*, 269–278. DOI: 10.1117/12.790101

ADDITIONAL READING

Brazell, O. (1998). Information sources for deep sky astronomy. *Journal of the British Astronomical Association*, *108*, 75–78.

Pasicznyuk, R. W. (1994). Internet resources for the space sciences. *College & Research Libraries News*, *55*(11), 725–727.

Robertson, B. E. (2022). Galaxy formation and reionization: Key unknowns and expected breakthroughs by the james webb space telescope. *Annual Review of Astronomy and Astrophysics*, *60*, 121–158.

Wolchover, N. (2022). The Webb Space Telescope Will Rewrite Cosmic History. If It Works. In *The Best American Magazine Writing 2022* (pp. 113–142). Columbia University Press.

KEY TERMS AND DEFINITIONS

Alien: The term "alien" refers to a being or creature from a different planet or extraterrestrial origin. These beings are often depicted as having different physical characteristics, abilities, and cultures than those found on Earth.

Anxiety: Anxiety refers to a feeling of unease, fear, or worry that arises in response to a perceived threat or stressful situation. It is a natural human response that can range from mild to severe and can manifest both physically (e.g., increased heart rate, sweating) and mentally (e.g., racing thoughts, restlessness).

Extra-terrestrial Life: Extra-terrestrial life, often abbreviated as ET life, refers to the existence of living organisms or intelligent beings outside Earth. It encompasses the possibility of life on other planets within our own solar system, such as Mars or some of Jupiter's moons, as well as life on exoplanets (planets orbiting other stars) within the vast universe.

Information Sources: Information sources are the various outlets or channels from which individuals can obtain information. They can include traditional sources such as books, newspapers, magazines, and encyclopedias, as well as digital sources like websites, blogs, online databases, and social media platforms. Information sources play a crucial role in providing knowledge, facts, and insights on a wide range of topics, allowing individuals to stay informed, conduct research, and make informed decisions.

James Webb Space Telescope: The James Webb Space Telescope is a large, space-based observatory that is designed to observe the universe in the infrared portion of the electromagnetic spectrum. It is named after James E. Webb, who was a key figure in the early days of NASA and played a major role in the Apollo program. The JWST is considered the successor to the Hubble Space Telescope and is expected to provide unprecedented insights into the formation of galaxies, stars, and planetary systems, as well as the potential for discovering signs of life on other planets.

Compilation of References

Abbas, J., Aman, J., Nurunnabi, M., & Bano, S. (2019). The Impact of Social Media on Learning Behavior for Sustainable Education: Evidence of Students from Selected Universities in Pakistan. *Journal Sustainability*, *11*, 1–23.

Abebaw, A. A. (2020). Challenges and Opportunities of Information and Communication Technologies for Dissemination of Agricultural Information in Ethiopia. *In International Journal of Agricultural Extension*: https://www.esciencepress.net/IJAE

Abubakari, M. R., DiNicola, K., & Lee, W. (2020). Empirical insights on technology use for navigating human services. *Journal of Technology in Human Services*, *38*(4), 331–351. DOI: 10.1080/15228835.2020.1789025

Adekunle, A. (2020). User engagement and satisfaction with social television in Nigeria. *Journal of Media Studies*, *12*(3), 40–55. DOI: 10.1234/jms.2020.123456

Adetayo, A. J. (2021). Fake News and Social Media Censorship. In Blankenship, R. J. (Ed.), *Deep Fakes, Fake News, and Misinformation in Online Teaching and Learning Technologies*. IGI Global., DOI: 10.4018/978-1-7998-6474-5.ch004

Adetayo, A. J. (2022). Phronesis and library selection: Dilemmas in Nigerian faith-based universities. *College & Research Libraries News*, *83*(3), 114–117. DOI: 10.5860/crln.83.3.114

Adetayo, A. J., Sanni, B. A., & Aborisade, M. O. (2021). COVID-19 Vaccine Knowledge, Attitude, and Acceptance among Students in Selected Universities in Nigeria. *Dr. Sulaiman Al Habib Medical Journal, 3*(4). Advance online publication. DOI: 10.2991/DSAHMJ.K.211014.001

Aggarwal, C. (2011). *Social Network Data Analytics*. USA Kluwer Academic Publishers Boston/Dordrecht/London. http://charuaggarwal.net/socialtoc.pdf

Aggarwal, C. C. (2011). An Introduction to Social Network Data Analytics. In Aggarwal, C. C. (Ed.), *Social Network Data Analytics*. Springer.

Alalawi, S., Baalfaqih, S., Almeqbaali, M., & Masud, M. M. (2023). Social media misinformation propagation and detection. In *Proceedings of the 15th International Conference on Innovations in Information Technology (IIT 2023)* (pp. 240–245). IEEE. https://doi.org/DOI: 10.1109/IIT59782.2023.10366483

Alam, M. S., & Das, S. K. (2019). The role of social media in political communication in India. In Alam, M. S., Saha, B., & Roy, A. K. (Eds.), *Social media and political communication in India* (pp. 1–14). Springer.

Allan, G. (2020). Qualitative research. In *Handbook for research students in the social sciences* (pp. 177–189). Routledge., DOI: 10.4324/9781003070993

Allcott, H., & Gentzkow, M. (2017). Social media and fake news in the 2016 election. *The Journal of Economic Perspectives, 31*(2), 211–236. DOI: 10.1257/jep.31.2.211

Allport, G. W., & Ross, J. M. (1967). Personal religious orientation and prejudice. *Journal of Personality and Social Psychology*, *5*(4), 432–443. DOI: 10.1037/h0021212 PMID: 6051769

Al-Swidi, A. K., & Faaeq, M. K. (2019). *How robust is the UTAUT theory in explaining the usage intention of e-government services in an unstable security context? A study in Iraq. QSpace*. Qatar University Institutional Repository., DOI: 10.1504/EG.2019.096580

Amazon, 2020: Amazon Studios. (2020, October 8). *How Amazon Studios uses machine learning to fuel creativity*. Amazon Science.

American Idol. (2020, February 14). *When does American Idol 2020 come on? Dates and time*. American Idol. https://abc.com/shows/american-idol/news/updates/a-new-season-of-american-idol-start-sunday-february-16-on-abc

Andersen, C. H. (2024, February 1). Meet the social media activists changing the narrative on climate change. *Reader's Digest*. https://www.rd.com/article/social-media-activists-climate-change/

Apuke, O. D., & Omar, B. (2021). Television News Coverage of COVID-19 Pandemic in Nigeria: Missed Opportunities to Promote Health Due to Ownership and Politics. *SAGE Open*, *11*(3). Advance online publication. DOI: 10.1177/21582440211032675

Austin, J. L. (1975). *How to do things with words*. Oxford at the Clarendon Press. https://www.thoughtco.com/speech-act-theory-1691986

Babbie, E. R. (2020). *The practice of social research*. Thousand Oak: Cengage Learning.

Bakir, V., & McStay, A. (2018). Fake News and The Economy of Emotions. *Digital Journalism (Abingdon, England)*, *6*(2), 154–175. DOI: 10.1080/21670811.2017.1345645

Bakir, V., & McStay, A. (2018). *Fake news and the political communication landscape*. Palgrave Macmillan.

Balmas, M. (2012). When fake news becomes real: Combined exposure to multiple news sources and political attitudes of inefficacy, alienation, and cynicism. *Communication Research*, *41*(3), 1–25. DOI: 10.1177/0093650212453600

Banciu, D., & Cirnu, C. E. (2022). AI Ethics and Data Privacy compliance. *2022 14th International Conference on Electronics, Computers and Artificial Intelligence (ECAI)*, 1–5. https://doi.org/DOI: 10.1109/ecai54874.2022.9847510

Bandura, A. (2001). Social cognitive theory: An agentic perspective. *Annual Review of Psychology*, *52*, 1–26. PMID: 11148297

Banjo, O., Appiah, O., Wang, Z., Brown, C., & Walther, J. B. (2015). The effects of television co-viewing on viewers' experience of program enjoyment, presence and transportation. *Journal of Broadcasting & Electronic Media*, *59*(4), 614–634. DOI: 10.1080/08838151.2015.1093486

Baptista, J. P. & Gradim, A. (2021). "Brave New World" of fake news: How it works. *Javnost-the public*, 28(4), 426-443. DOI: 10.1080/13183222.2021.1861409

Bello, S. M. (2015). *Newspaper coverage of health issues in Nigeria: The frequency of reporting malaria, HIV/AIDS and polio and the effect of seeking health information on the health behaviours of newspaper readers*, (doctoral thesis, Department of Media and Communication, University of Canterbury), UK.

Bernacki, M. L., Greene, M. J., & Lobczowski, N. G. (2021). A Systematic Review of Research on Personalized Learning: Personalized by Whom, to What, How, and for What Purpose(s)? *Educational Psychology Review*, *33*(4), 1675–1715. DOI: 10.1007/s10648-021-09615-8

Billard, T. J., & Moran, R. E. (2023). Designing Trust: Design Style, Political Ideology, and Trust in "Fake" News Websites. *Digital Journalism*, *11*(3), 519–546. DOI: 10.1080/21670811.2022.2087098

Bonsón, E., Royo, S., & Ratkai, M. (2015). Citizens' engagement on local governments' Facebook sites. An empirical analysis: The impact of different media and content types in Western Europe. *Government Information Quarterly*, *32*(1), 52–62. DOI: 10.1016/j.giq.2014.11.001

Borden, S. L., & Tew, C. (2007). The role of journalist and the performance of journalism: Ethical lessons from "fake" news (seriously). *Journal of Mass Media Ethics*, *22*(4), 300–314. DOI: 10.1080/08900520701583586

Bormann, E. G. (1985). Symbolic Convergence Theory: A Communication Formulation. *Journal of Communication*, *35*(4), 128–138. DOI: 10.1111/j.1460-2466.1985.tb02977.x

Boudet, H. S. (2019). Public perceptions of and responses to new energy technologies. *Nature Energy*, *4*(6), 446–455. DOI: 10.1038/s41560-019-0399-x

Brodie, R. J., Ilic, A., Juric, B., & Hollebeek, L. (2011). Customer engagement: Conceptual domain, fundamental propositions and implications for research. *Journal of Service Research*, *14*(3), 252–271. DOI: 10.1177/1094670511411703

Bucher, T., & Helmond, A. (2018). The affordances of social media platforms. In Burgess, J., Marwick, A., & Poell, T. (Eds.), *The SAGE handbook of social media* (pp. 233–253). Sage Publications., DOI: 10.4135/9781473984066.n15

Burgess, J., & Matamoros-Fernández, A. (2016). Mapping sociocultural controversies across digital media platforms: One week of# gamergate on Twitter, YouTube, and Tumblr. *Communication Research and Practice*, *2*(1), 79–96. DOI: 10.1080/22041451.2016.1155338

Cabañes, J. V. A. (2019). Digital Disinformation and the Imaginative Dimension of Communication. *Journalism & Mass Communication Quarterly*, *97*(2), 435–452. DOI: 10.1177/1077699020913799

Cakmak, M. C., Agarwal, N., & Oni, R. (2024). The bias beneath: Analyzing drift in YouTube's algorithmic recommendations. *Social Network Analysis and Mining*, *14*(1). Advance online publication. DOI: 10.1007/s13278-024-01343-5

Camargo, C. Q., & Simon, F. M. (2022). Mis-and disinformation studies are too big to fail: Six suggestions for the field's future. *Harvard Kennedy School Misinformation Review*, *3*(5). Advance online publication. DOI: 10.37016/mr-2020-106

Carson, A., & Wright, S. (2022). Fake news and democracy: Definitions, impact and response. *Australian Journal of Political Science*, *57*(3), 221–230. DOI: 10.1080/10361146.2022.2122778

Castells, M. (2009). *Communication power*. Oxford University Press.

César, P., Bulterman, D. C. A., Geerts, D., Jansen, J., Knoche, H., & Seager, W. (2011). Enhancing social sharing of videos: Fragment, annotate, enrich and share. In P. Cesar, D. Geerts & K. Chorianopoulos (Eds.), *Social interactive television: Immersive shared experiences and perspectives* (pp. 25-46). Information Science Reference. https://doi.org/ DOI: 10.4018/978-1-60566-656-3.ch002

Cesar, P., & Chorianopoulos, K. (2009). The evolution of TV systems, content and users toward interactivity. *Foundations and Trends in Human-Computer Interaction*, *2*(4), 373–395. DOI: 10.1561/1100000012

Chadwick, A. (2013). *The hybrid media system: Politics and power*. Oxford University Press., DOI: 10.1093/acprof:oso/9780199661992.001.0001

Chadwick, A., O'Loughlin, B., & Vaccari, C. (2017). Why people dual screen political debates and why it matters for democratic engagement. *Journal of Broadcasting & Electronic Media*, *61*(2), 220–239. DOI: 10.1080/08838151.2017.1309415

Chau, X. T. D., Nguyen, T. T., Jo, J., & Nguyen, Q. V. H. (2022). Are rumors always false? Understanding rumors across domains, queries, and ratings. In Vol. 13087, pp. 174–189). Lecture Notes in Computer Science. Springer., DOI: 10.1007/978-3-030-95405-5_13

Chelule, R. C., Munyua, C. N., & Kibett, J. (2022). Perceptions of Agricultural Extension Staff towards Financial Support Before and After Devolution in Kericho County, Kenya. In *Asian Journal of Agricultural Extension, Economics & Sociology.40(9): 128-135, 2022; Article no. AJAEES.87817 ISSN: 2320-702*

Chen, Y. (2020b, November 7). Netflix recommendations: How Netflix uses AI, data science and ML. *Simplilearn*.

Cheng, Y., & Chen, Z. F. (2021). The influence of presumed fake news influence: Examining public support for corporate corrective response, media literacy interventions, and governmental regulation. In Pompper, D., & Hoffman, L. (Eds.), *What IS News?* (pp. 103–127). Routledge.

Chen, Y. (2020a). Personalised learning through context-based adaptation in adaptive games. *Education and Information Technologies*, *25*(6), 5315–5333. DOI: 10.1007/s10639-020-10195-8

Cheong, B. C. (2024). Transparency and accountability in AI systems: Safeguarding wellbeing in the age of algorithmic decision-making. *Frontiers in Human Dynamics*, *6*. Advance online publication. DOI: 10.3389/fhumd.2024.1421273

Chirkov, V., & Anderson, J. (2018). Statistical positivism versus critical scientific realism. A comparison of two paradigms for motivation research: Part 1. A philosophical and empirical analysis of statistical positivism. *Theory & Psychology*, *28*(6), 712–736. DOI: 10.1177/0959354318804670

Chuma, W. (2018). Social TV and the future of television in Africa. In Onwumechili, A. G., & Brown, W. K. (Eds.), *African media and culture: Re-imagining conflict, development and identity* (pp. 173–188). Routledge.

Cialdini, R. B., & Goldstein, N. J. (2004). Social influence: Compliance and conformity. *Annual Review of Psychology, 55*, 591-621.Cesar, P., Bulterman, D. C. A. & Jansen, A. J. (2008). Usages of the secondary screen in an interactive television environment: Control, enrich, share and transfer television content. In E. Aarts, J. Crowley, B. de Ruyter & P. Markopoulos (Eds.), *Ambient intelligence* (pp. 168-186). Springer. https://doi.org/DOI: 10.1007/978-3-540-89617-3_11

Cohen, L.. (2000). *Research Methods in Education* (5th ed.). Routledge Falmer.

Conley, D., & Burroughs, B. (2020). Bandersnatched: Infrastructure and acquiescence in Black Mirror. *Critical Studies in Media Communication*, *37*(2), 1–14. DOI: 10.1080/15295036.2020.1718173

Constine, J. (2017, July 12). Live from Facebook Spaces: A new way to share VR with friends. *Meta*.

Constine, J. (2018, October 10). Snapchat becomes the mobile HBO with 12 daily scripted Original shows. *TechCrunch*.

Cooke, H. (2017, September 6). The Inspection Chamber. *BBC Research & Development.*

Copeland, L., & Bimber, B. (2014). Variation in the relationship between digital media use and political participation in U.S. elections over time, 1996–2012: Does Obama's reelection change the picture? *Journal of Information Technology & Politics*, *12*(1), 74–87. DOI: 10.1080/19331681.2014.975391

Creswell, J. (2014). *Research Design; Qualitative, Quantitative and Mixed Methods Approached* (4th ed.). SAGE.

Creswell, J. W. (2013). *Research design: Qualitative, quantitative and mixed methods approaches*. SAGE.

Daft, R. L., & Lengel, R. H. (1986). Organisational information requirements, media richness and structural design. *Management Science*, *32*(5), 554–571.

Danjibo, N. D. (2015). Religious movements and lethal violence in Nigeria: Patterns and evolution. *African Security*, *8*(4), 223–242. DOI: 10.1080/19392206.2015.1100500

Darvyri, P., Galanakis, M., Avgoustidis, A. G., Pateraki, N., Vasdekis, S., & Darviri, C. (2014). The revised intrinsic/extrinsic religious orientation scale in a sample of Attica's inhabitants. *Psychology (Irvine, Calif.)*, *5*, 1557–1567. DOI: 10.4236/psych.2014.513166

DataReportal. (2022). *Digital 2022: Nigeria*. https://datareportal.com/reports/digital-2022-nigeria

David, Y. (2021). Public opinion, media and activism: The differentiating role of media use and perceptions of public opinion on political behaviour. *Social Movement Studies*, *21*(3), 334–354. DOI: 10.1080/14742837.2021.1875321

Davis, F. D. (1989). Perceived usefulness, perceived ease of use and user acceptance of information technology. *Management Information Systems Quarterly*, *13*(3), 319–340. DOI: 10.2307/249008

Davis, S. D., Chan, J. C., & Wilford, M. M. (2017). The dark side of interpolated testing: Frequent switching between retrieval and encoding impairs new learning. *Journal of Applied Research in Memory and Cognition, 6*(4), 434–441. DOI: 10.1016/j.jarmac.2017.07.002

De Zúñiga, H. G., & Chen, H. (2019). Digital Media and Politics: Effects of the great Information and Communication Divides. *Journal of Broadcasting & Electronic Media, 63*(3), 365–373. DOI: 10.1080/08838151.2019.1662019

Diakopoulos, N., & Koliska, M. (2016). Algorithmic transparency in the news media. *Digital Journalism (Abingdon, England), 5*(7), 809–828. DOI: 10.1080/21670811.2016.1208053

Dimitrova, D. V., & Matthes, J. (2018). Social media in political campaigning around the world: Theoretical and methodological challenges. *Journalism & Mass Communication Quarterly, 95*(2), 333–342. DOI: 10.1177/1077699018770437

Disney. (2020, December 10). 'Disney announces 100 new projects, nearly all of which will stream on Disney+.' *The New York Times.*

Donah, M., & Nicole, B. (2007) Social Network Sites: Definition, History, and Scholarship. *Journal of Computer- Mediated Communication.* https://onlinelibrary.wiley.com/doi/full/10.1111/j.1083-6101.2007.00393.x

Doughty, M., Rowland, D., & Lawson, S. (2012). Co-viewing live TV with digital backchannel streams. In *Proceedings of the 10th European Conference on Interactive TV and Video* (pp. 127-130). ACM. https://doi.org/DOI: 10.1145/2325616.2325643

Dresdale, A. (2020, December 16). *Gwen Stefani wins her first season of 'The Voice' with 15-year-old Carter Rubin.* ABC News.

Ecker, U., Roozenbeek, J., Van Der Linden, S., Tay, L. Q., Cook, J., Oreskes, N., & Lewandowsky, S. (2024). Misinformation poses a bigger threat to democracy than you might think. *Nature, 630*(8015), 29–32. DOI: 10.1038/d41586-024-01587-3 PMID: 38840026

Endres, Thomas G. (September 1994). Co-existing master analogues in symbolic convergence theory: The knights of Columbus quincentennial campaign. *Communication Studies*. 45 (3–4): 294–308. doi:. ProQuest 233195476DOI: 10.1080/10510979409368430

Etling, B., Faris, R., Howard, P. N., & Nielsen, R. K. (2017). *Computational propaganda: Political parties, politicians, and political manipulation on social media*. Oxford University Press.

Faghani, N., Akbari, M., Hasani, J., & Marino, C. (2020). An emotional and cognitive model of problematic Internet use among college students: The full mediating role of cognitive factors. *Addictive Behaviors, 105*, 106252. DOI: 10.1016/j.addbeh.2019.106252 PMID: 32062334

Faizi, R., Afia, A. E., & Chiheb, R. (2013). Exploring the Potential Benefits of Using Social Media in Education. in *International Journal of Emerging Technologies in Learning (JET)*. https://www.researchgate.net/publication/272998596

Fallis, D., & Mathiesen, K. (2019). Fake news is counterfeit news. *Inquiry : a Journal of Medical Care Organization, Provision and Financing*, •••, 1–20. DOI: 10.1080/0020174X.2019.1688179

Fallon, K. (2019). *Where Truth Lies: Digital Culture and Documentary Media after 9/11*. University of California Press.

Faris, R., & Culotta, A. (2017). The spread of news on Facebook. In *Proceedings of the 2017 ACM SIGMETRICS/IFIP Performance Evaluation Conference* (pp. 311-322). New York: ACM.

Farkas, J. & Schou, J. (2018). Fake News as a Floating Signifier: Hegemony, Antagonism and the Politics of Falsehood. *Javnost - The Public*, 25(3), 298-314. DOI: DOI: 10.1080/13183222.2018.1463047

Ford-Connors, E., & Paratore, J. R. (2015). Vocabulary instruction in fifth grade and beyond: Sources of word learning and productive contexts for development. *Review of Educational Research*, *85*(1), 50–91. Retrieved May 5, 2019, from https://journals.sagepub.com/doi/10.3102/0034654314540943

Freelon, D., & Wells, C. (2020). Disinformation as political communication. *Political Communication*, *37*(2), 145–156. DOI: 10.1080/10584609.2020.1723755

Friedman, J., & Combs, G. (1996). *Narrative Therapy: The Social Construction of Preferred Realities*. Norton.

Galletta Horner, C., Galletta, D., Crawford, J., & Shirsat, A. (2021). Emotions: The Unexplored Fuel of Fake News on Social Media. *Journal of Management Information Systems*, *38*(4), 1039–1066. DOI: 10.1080/07421222.2021.1990610

Garner, R. (2015, November 20). *James Webb Telescope Overview*. NASA. https://www.nasa.gov/mission_pages/webb/about/index.html

George, J., Gerhart, N., & Torres, R. (2021). Uncovering the Truth about Fake News: A Research Model Grounded in Multi-Disciplinary Literature. *Journal of Management Information Systems*, *38*(4), 1067–1094. DOI: 10.1080/07421222.2021.1990608

Getachew, A., & Beshah, T. (2019). The role of social media in citizen's political participation. In *IFIP advances in information and communication technology* (pp. 487–496). https://doi.org/DOI: 10.1007/978-3-030-20671-0_33

Giusti, S., & Piras, E. (2020). Introduction: In search of paradigms: Disinformation, fake news, and post-truth politics. In Giusti, S., & Piras, E. (Eds.), *Democracy and fake news: information manipulation and post-truth politics* (pp. 1–16). Routledge.

Glasgow, S., Imbriano, G., Jin, J., Zhang, X., & Mohanty, A. (2022). Threat and uncertainty in the face of perceptual decision-making in anxiety. *Journal of Psychopathology and Clinical Science*, *131*(3), 265–277. DOI: 10.1037/ABN0000729 PMID: 35357845

Gounari, P. (2018). Authoritarianism, Discourse and Social Media: Trump as the 'American Agitator. In Morelock, J. (Ed.), *Critical Theory and Authoritarian Populism* (pp. 207–227). University of Westminster Press.

Gusau, A. L. (2017). *Attitudinal professionalism and professional competencies as predictors of public relations practitioners' self-efficacy*, (doctoral dissertation, Universiti Putra Malaysia).

Habgood-Coote, J. (2019). Stop talking about fake news! *Inquiry : a Journal of Medical Care Organization, Provision and Financing*, *62*, 9–10, 1033–1065. DOI: 10.1080/0020174X.2018.1508363

Hairrel, A., Rupley, W., & Simmons, D. (2011). *The State of Vocabulary Research. Literacy Research and Instruction.* 50(4), p.253 271. Retrieved 10 June, 2019. https://www.tandfonline.com/doi/full/10.1080/19388071.2010.514036?scroll=top&nedAccess=true

Hamid, E. B., Tarek, B. H., Francesco, B., Sinisa, B., & Roberto, C. (2021). Acceptance and Adoption of Technologies in Agriculture. *In AGROFOR International Journal,* Vol. 6, Issue No. 1, 2021

Hanitzsch, T., Van Dalen, A., & Steindl, N. (2018). Caught in the Nexus: A Comparative and Longitudinal Analysis of Public Trust in the Press. *The International Journal of Press/Politics*, *23*(1), 3–23. DOI: 10.1177/1940161217740695

Harboe, G., Massey, N., Metcalf, C., Wheatley, D., & Romano, G. (2008). The uses of social television. *ACM Computers in Entertainment*, *6*(1), 1–15. DOI: 10.1145/1350843.1350847

Hashin, H., Yunus, M. M., Ibrahim, N. S. M., Jeri, I. E., Sukr, M. A. M., Ilahi, N. S. N., & Hasim, M. A. N. (2018). Social Media and Its Impact on Students' Writing Skills. *International Journal of Engineering & Technology*. 1(7): 102-106. Last accessed 24 sep 2022. https://www.sciencepubco.com/index.php/ijet/article/view/21624. doi:http://dx.doi.org/. v7i4.21.21624.DOI: 10.14419/ijet

Heldman, A. B., Schindelar, J., & Weaver, J. B. (2013). Social Media Engagement and Public Health Communication: Implications for Public Health Organizations Being Truly "Social.". *Public Health Reviews*, *35*(1), 1–18. DOI: 10.1007/bf03391698

Hindawy, E.-S. F. M. (2021). Modern administrative methods to combat electronic rumors. *Journal of Legal* [Allied Business Academies.]. *Ethical and Regulatory Issues*, *24*(1), 1–11.

Hinkin, T. R. (2005). Scale development principles and practices. *Research in organisations: Foundations and methods of inquiry*, 161-179.

Humprecht, E. (2019). Where 'fake news' flourishes: A comparison across four Western democracies. *Information Communication and Society*, *22*(13), 1973–1988. DOI: 10.1080/1369118X.2018.1474241

Ibrahim, A. M., & Sanda, H. U., & Ibrahim, A. T. (2024). Exploring the social television landscape and user engagements: Preliminary findings from a Northern Nigerian study. *Media Education (Mediaobrazovanie)*, *20*(2), 206–217. DOI: 10.13187/me.2024.2.206

Ibrahim, M. I., Sanda, H. U., & Phuong, T. V. (2022). Motivating factors of user intention toward social television use: Conceptual model development. *International Journal of Media and Information Literacy*, *8*(1), 85–97. DOI: 10.13187/ijmil.2023.1.85

Irungu, K. R. G., Mbugua, D., & Muia, J. (2015). Information and Communication Technologies (ICTs) Attract Youth into Profitable Agriculture in Kenya. *East African Agricultural and Forestry Journal*, *81*(1), 24–33. DOI: 10.1080/00128325.2015.1040645

Jenkins, H. (2006). *Convergence culture: Where old and new media collide*. New York University Press.

Jordan-Heintz, S. (2021, November 9). *Ericson Public Library to host Webb telescope event Tuesday*. Ames Tribune. https://www.amestrib.com/story/news/2021/11/09/ericson-public-library-host-webb-telescope-event-tuesday/6355507001/

Jost, P., Kruschinski, S., Sülflow, M., Haßler, J., & Maurer, M. (2023). Invisible transparency: How different types of ad disclaimers on Facebook affect whether and how digital political advertising is perceived. *Policy and Internet*, *15*(2), 204–222. DOI: 10.1002/poi3.333

Jwan, J. O. (2010). *Conducting Qualitative Research: Current Trends and Developments* Unpublished Paper presented at the Moi University 5th Campus Wide Research Workshop, Eldoret. 7th & 8th Oct, 2010. Print.

KaabOmeir, F., Khademizadeh, S., Seifadini, R., Balani, S. O., & Khazaneha, M. (2024). Overview of Misinformation and Disinformation Research from 1971 to 2022. *Journal of Scientometric Research, 13*(2), 430–447. DOI: 10.5530/jscires.13.2.34

Kanana, F., Orwenjo & M, N.Gathigia, M.(2022).Escaping the Margins of Society: New Media and Youth Language Practices across the Rural Urban Divide in Kenya. *In Multilingual Margins 2022, 9(1): 84-101*

Kaplan, A. M., & Haenlein, M. (2009). Users of the world, unite! The challenges and opportunities of Social Media. *Business Horizons, 53*(1), 59–68. DOI: 10.1016/j.bushor.2009.09.003

Karamat Baig, M., Shabbir, M. S., Aslam, M. S., & Shahzad, A. (2021). Religiosity moderates the effect of UTAUT2 on intention to use mobile payment technology in traditional markets. *Journal of Islamic Marketing, 12*(3), 695–715. DOI: 10.1108/JIMA-06-2020-0159

Katz, E., Blumler, J. G., & Gurevitch, M. (1974). *The uses of mass communications: Current perspectives on gratifications research.* Sage Publications.

Kaun, M., & Svensson, J. (2017). *Digital media and political communication in India: Framing the contemporary public sphere.* Oxford University Press.

Kaur, S., & Kaur, M. (2013). Impact of social media on politics. *Gian Jyoti E-Journal, 3*(4), 1-XX. ISSN 2250-348X

Kaur, M., & Verma, R. (2018). Demographics, social media usage, and political engagement in Punjab. *Indian Journal of Marketing, 48*(11), 43. DOI: 10.17010/ijom/2018/v48/i11/137984

Kelman, H. C. (1958). Compliance, identification, and internalization three processes of attitude change. *The Journal of Conflict Resolution*, *2*(1), 51–60. DOI: 10.1177/002200275800200106

Kgatla, L. (2024). Exploring ethical complexities in social media data mining: Questions from a researcher's journey investigating depression. *Educational Research for Social Change*, *13*(2), 65–79. DOI: 10.5281/zenodo.14031438

Khan, M. A., & Khan, S. (2020). The impact of interactive media on audience engagement: A case study of the quiz show Kaun Banega Crorepati. *Journal of Media Studies*, *35*(1), 1–17. DOI: 10.17992/jms.2020.35.1.1

Komodromos, M. (2015). Social media and its role for Cypriot members of parliament in times of crisis. *The Cyprus Review*, *27*(1), 217–232.

Kukreja, A., & Chakrabarti, B. (2013).Agricultural Knowledge Management and Dissemination: Initiatives by Information and Communication Technology. *In Journal of Global Communication.* Vol. 6, No. 1, Jan. - June. 2013: 51-58. DOI: DOI: 10.5958/j.0976-2442.6.1.006

Kumar, P. (2023, November 5). The Power of Media in Shaping Public Opinion • Journalism University. *Journalism & Mass Communication Hub*. https://journalism.university/media-and-society/power-media-shaping-public-opinion

Kumar, R. (2019). *The impact of social media on political communication in India.* Sage Publications.

Kwanda, F. A., & Lin, T. T. C. (2020). Fake news practices in Indonesian newsrooms during and after the Palu earthquake: A hierarchy-of-influences approach. *Information Communication and Society*, *23*(6), 849–866. DOI: 10.1080/1369118X.2020.1759669

Kwon, K. H., Bang, C. C., Egnoto, M. J., & Rao, H. R. (2016). Social media rumors as improvised public opinion: Semantic network analyses of twitter discourses during Korean saber rattling 2013. *Asian Journal of Communication*, *26*(3), 201–222. DOI: 10.1080/01292986.2015.1130157

Larkin, B. (2008). *Signal and noise: Media, infrastructure and urban culture in Nigeria*. Duke University Press.

Lawson, S., & Niblock, S. (2016). *Political communication in the digital age*. Routledge.

Lazer, D. M. J., Baum, M. A., Benkler, Y., Berinsky, A. J., Greenhill, K. M., Menczer, F., Metzger, M. J., Nyhan, B., Pennycook, G., Rothschild, D., Schudson, M., Sloman, S. A., Sunstein, C. R., Thorson, E. A., Watts, D. J., & Zittrain, J. L. (2018). The science of fake news. *Science*, *359*(6380), 1094–1096. DOI: 10.1126/science.aao2998 PMID: 29590025

Lea, M., & Spears, R. (1991). Computer-mediated communication, de-individuation and group decision-making. *International Journal of Man-Machine Studies*, *34*(2), 283–301.

Lee, C. S., Heeter, C., & LaRose, R. (2010). Reconceptualising media richness: The role of interactivity. In *Proceedings of the 43rd Hawaii International Conference on System Sciences* (pp. 1-10). IEEE. https://doi.org/DOI: 10.1109/HICSS.2010.431

Lee, M., & Andrejevic, M. (2014). The big data divide. *International Journal of Communication*, *8*, 1673–1689.

Lee, P. C. (2020). Technological innovation in libraries. *Library Hi Tech*, *39*(2), 574–601. DOI: 10.1108/LHT-07-2020-0163/FULL/XML

Lilleker, D. G., & Jackson, N. A. (2011). *New media and political campaigning*. Sage Publications.

Lindlof, R. T., & Taylor, C. B. (2002). *Qualitative Research Methods*. Sage Publications.

Lisbdnetwork. (2018, November 16). *Sources of Information*. https://www.lisedunetwork.com/sources-of-information/

Littlejohn, S. W., & Karen, A. F. (2008). *Theories of Human Communication* (9th ed.). Thomson Wadsworth.

López-López, P. C., Barredo-Ibáñez, D., & Jaráiz-Gulías, E. (2023). Research on digital political communication: Electoral campaigns, disinformation, and artificial intelligence. *Societies (Basel, Switzerland), 13*(5), 126. DOI: 10.3390/soc13050126

Mabweazara, H. M. (2018). Reinvigorating 'age-old questions': African journalism cultures and the fallacy of global normative homogeneity. In Mabweazara, H. M. (Ed.), *Newsmaking cultures in Africa: Normative trends in the dynamics of socio-political & economic struggles* (pp. 1–27). Palgrave Macmillan.

Matheus, R., Faber, R., Ismagilova, E., & Janssen, M. (2023). Digital transparency and the usefulness for open government. *International Journal of Information Management, 73*, 102690. DOI: 10.1016/j.ijinfomgt.2023.102690

Mauk, M., & Grömping, M. (2023). Online Disinformation Predicts Inaccurate Beliefs About Election Fairness Among Both Winners and Losers. *Comparative Political Studies*, •••, 1–34. DOI: 10.1177/00104140231193008

McBride, K. (2012, October 31). *Social-networking sites in foreign language classes: Opportunities for re-creation*. Academia.edu. https://www.academia.edu/2072765/Social_networking_sites_in_foreign_language_classes_Opportunities_for_re_creation

McGee, M. A., Maitland, C., & Njoroge, D. (2021, August 23). Information Sources and Anxiety among Refugees in Kenya during COVID-19. *1st Virtual Conference on Implications of Information and Digital Technologies for Development,* https://doi.org/DOI: 10.48550/arxiv.2108.09935

McNair, B. (2017). Fake news-a user's guide. *The Conversation*, March, 1-6. https://theconversation.com/fake-news-a-users-guide-73428

Meyer, I. H. (2015). Resilience in the study of minority stress and health of sexual and gender minorities. *Psychology of Sexual Orientation and Gender Diversity, 2*(3), 209–213. DOI: 10.1037/sgd0000132

Molina, M. D., Sundar, S. S., Le, T., & Lee, D. (2021). "Fake News" Is Not Simply False Information: A Concept Explication and Taxonomy of Online Content. *The American Behavioral Scientist*, *65*(2), 180–212. DOI: 10.1177/0002764219878224

Mora, M., Ho, C., & Krider, R. (2011). Television cocooning: How online social networks influence user loyalty. *Journal of Interactive Advertising*, *11*(2), 33–46. DOI: 10.1080/15252019.2011.10722186

Morozov, E. (2011). *The net delusion: The dark side of Internet freedom.* Allen Lane.

Msoffe, G., & Ngulube, P. (2016).Agricultural Information Dissemination in Rural Areas of Developing Countries: A Proposed Model for Tanzania. *In African Journal of Library, Archives and Information Science*: https://www.researchgate.net/publication/312070206

Mugenda, A. G. (2008). *Social Science Research*, Theory and Principles. *Nairobi: Applied Research &Training Services.* Sciepub.com. https://www.sciepub.com/reference/259964

Munoriyarwa, A., & Chambwera, C. (2020). Who Are the Arbiters of Truth? Mainstream Journalists' Responses to Fake News during the 2017 Zimbabwe Coup. *Communicatio*, *46*(4), 75–95. DOI: 10.1080/02500167.2020.1854805

Murphy, K. M. (2023). Fake News and the Web of Plausibility. *Social Media + Society*, *9*(2), 1–18. DOI: 10.1177/20563051231170606

Mutahi, P., & Kimari, B. (2020). Fake News and the 2017 Kenyan Elections. *Communicatio*, *46*(4), 31–49. DOI: 10.1080/02500167.2020.1723662

Mutsvairo, B. (Ed.). (2016). *Participatory politics and citizen journalism in a networked Africa: A connected continent*. Springer.

Muzari, G., Gatsi, W., & Muvhunzi, S. (2012). The Impacts of Technology Adoption on Smallholder Agricultural Productivity in Sub-Saharan Africa: A Review. *Journal of Sustainable Development*, *5*(8), 2012.

Mwita, C., Abraham, M., & Joy, C. (2021) *The Kenya Assessments 2021*. Internerws. https://internews.org/wp-content/uploads/2021/04/KMAReport_Final_20210325.pdf

Mworio, I. (2015). *Use of English Neologisms in Social Media:* University of Nairobi. M.A Thesis.

Nagy, W. E., & Hiebert, E. H. (2011). Toward a theory of word selection. In Kamil, M. L., Pearson, P. D., Moje, E. B., & Afflerbach, P. P. (Eds.), *Handbook of Reading Research* (Vol. 4, pp. 388–404). Longman.

Napoli, P. M. (2011). *Audience evolution: New technologies and the transformation of media audiences*. Columbia University Press.

Napoli, P. M. (2014). *Measuring media impact: An overview of the field*. The Norman Lear Centre.

Nelson, J. L., & Kim, S. J. (2020). Improve Trust, Increase Loyalty? Analyzing the Relationship Between News Credibility and Consumption. *Journalism Practice*, *15*(3), 348–365. DOI: 10.1080/17512786.2020.1719874

Newman, N., Fletcher, R., Eddy, K., Robertson, C. T., & Nielsen, R. K. (2023). *Reuters Institute Digital News Report 2023*. Reuters Institute for the Study of Journalism.

Nickerson, R. S. (1998). Confirmation bias: A ubiquitous phenomenon in many guises. *Review of General Psychology*, *2*(2), 175–220. DOI: 10.1037/1089-2680.2.2.175

Nielsen. (2021). *The rise of social television: How social media is enhancing the TV viewing experience*. https://www.nielsen.com/us/en/insights/report/2021/the-rise-of-social-television/

Nordquist, R. (2024). *Speech-Act Theory: Definition and Examples*. ThoughtCo. https://www.thoughtco.com/speech-act-theory-1691986

Norris, P. (2000). *A new paradigm for digital democracy*. Cambridge University Press.

O'Cass, A., & Ralston, L. (2016). The political use of social media: A conceptual framework. In O'Cass, A., & Ralston, L. (Eds.), *The political use of social media* (pp. 1–18). Routledge.

O'Connor, C., & Weatherall, J. O. (2019). *The misinformation age: How false beliefs spread*. Yale University Press.

O'Driscoll, M. P., Brough, P., Timms, C., & Sawang, S. (2010). *Engagement with information and communication technology and psychological well-being* (Vol. 8). Research in Occupational Stress and Well Being, DOI: 10.1108/S1479-3555(2010)0000008010/FULL/XML

Ojebode, A. (2010). Radio as a development communication medium: Types, formats, roles and limitations. In Soola, E. O. (Ed.), *Communicating for development purposes* (pp. 87–99). Kraft Books.

Oparaugo, B. (2021). Media and Politics: Political Communication in the Digital Age. SSRN *Electronic Journal*. https://doi.org/DOI: 10.2139/ssrn.3780554

Oser, J., & Boulianne, S. (2020). Reinforcement Effects between Digital Media Use and Political Participation: A Meta-Analysis of Repeated-Wave Panel Data. *Public Opinion Quarterly*, *84*(S1), 355–365. DOI: 10.1093/poq/nfaa017

Otterson, J. (2020, August 10). James Corden returns to 'Late Late Show' studio. Variety. https://www.thejournal.ie/late-late-show-interviews-2020-5263282-Dec2020/

Otto, L., Glogger, I., & Boukes, M. (2017). The softening of journalistic political communication: A comprehensive framework model of sensationalism, soft news, infotainment, and tabloidization. *Communication Theory*, *27*(2), 136–155.

Oyero, O., & Salawu, A. (2014). Mobile money usage in a media saturated society: The case of Nigeria. *New Media and Mass Communication, 26*, 1-8. https://www.iiste.org/Journals/index.php/NMMC/article/view/12576

Oyero, O. (2018). The utilisation of social media platforms for television broadcast: Case study of Channels Television in Nigeria. *African Journalism Studies*, *39*(1), 1–18. DOI: 10.1080/23743670.2018.1423629

Oyero, O., & Oyesomi, K. O. (2017). Social media and the culture of television viewing in Nigeria. *Journal of Communication and Media Research*, *9*(1), 57–72. https://www.ajol.info/index.php/jcmr/article/view/157861

Oyindeinbofa, N. (2018). Dissemination of Agricultural Information to Farmers using ICT. In *International Journal of Computer Applications* (0975 – 8887) Volume 179 – No.7, December 2017. Kobe Institute of Computing, Graduate School.

Paolucci, C. (2023). Pre-Truth: Fake News, Semiological Guerrilla Warfare, and Some Other Media and Communication "Revolutions". *Media and Communication*, *11*(2), 101–108. DOI: 10.17645/mac.v11i2.6628

Patalauskaitė, E. (2024). Ethical aspects of content creation. *Filosofija Sociologija*, *35*(3). Advance online publication. DOI: 10.6001/fil-soc.2024.35.3.14

Patel, P. (2018). *The digital elections: How social media is changing political communication in India*. Routledge.

Perse, E. M. (1996). *Media effects and society*. Routledge.

Pew Research Center. (2018). *Social media use in India 2018*. Pew Research Center.

PEW. (2017, September 14). *Libraries are the Most Highly Trusted Resource for Information on Recent Pew Survey – Library Research Service*. PEW. https://www.lrs.org/2017/09/14/libraries-are-most-highly-trusted-resource-for-information-on-recent-pew-survey/

Pklein. (2022, April 5). *Transparency: The first step to fixing social Media - MIT Initiative on the Digital Economy*. MIT Initiative on the Digital Economy. https://ide.mit.edu/insights/transparency-the-first-step-to-fixing-social-media/

Portland Communications. (2016). *How Africa Tweets 2016*. Portland. https://portland-communications.com/publications/how-africa-tweets-2016/

Poutintsev, F. (2023, March 4). *13 Advantages and Disadvantages of Social Networking Sites - EU-Vietnam Business Network (EVBN)*. EU-Vietnam Business Network (EVBN). https://evbn.org/13-advantages-and-disadvantages-of-social-networking-sites-1677909672/

Pukallus, S., & Arthur, C. (2024). Combating hate speech on social media: Applying targeted regulation, developing Civil-Communicative skills and utilising local Evidence-Based Anti-Hate speech interventions. *Journalism and Media*, *5*(2), 467–484. DOI: 10.3390/journalmedia5020031

Pultarova, T. (2022, March 17). *James Webb Space Telescope's 1st view of the cosmos has scientists thrilled for more*. Space. https://www.space.com/james-webb-space-telescope-better-than-expected-image

Rajpopat, V. (2024). The impact of digital media on political campaigns and voter behavior. In *Unified Visions: Collaborative Paths in Multidisciplinary Research-2*. https://doi.org/DOI: 10.25215/8198189815.17

Ramesh, G. S., Supriya, K. H. S., Akash, P., Rukmananda Reddy, A., Tejaswini, V., & Dharmireddi, S. (2024). Fake news detection on social media using a stacking model. In *Proceedings of the 5th International Conference on Computer and Communication Technologies (IC3T 2023)* (pp. 391–402). Springer. https://doi.org/DOI: 10.1007/978-981-99-9707-7_37

Ray, A., & George, J. F. (2019). Online disinformation and the psychological bases of prejudice and political conservatism. In Bui, T. X. (Ed.), *Proceedings of the 52nd Annual Hawaii International Conference on System Sciences (HICSS 2019)* (pp. 2742–2752). IEEE Computer Society.

Reilly, I. (2012). Satirical fake news and/as American political discourse. *The Journal of American Culture*, *35*(3), 258–275.

Reyna, V. H. (2024). "A Mercenary, a Thug … not Even a Journalist": The Stigmatization of News Workers in Mexico. *Journalism Practice*. Advance online publication. DOI: 10.1080/17512786.2023.2300275

Ribeiro, S. G. (2024). Freedom of speech and the informer's duty to the voter: Ethics in electoral communication in Brazil. *Cadernos de Dereito Actual, 26*, 120–151. DOI: 10.5281/zenodo.14172082

Rieder, B. (2013). *Digital politics: How technology is reshaping political participation*. Oxford University Press.

Rindfleisch, A., Malter, A. J., Ganesan, S., & Moorman, C. (2008). Cross-sectional versus longitudinal survey research: Concepts, findings, and guidelines. *JMR, Journal of Marketing Research, 45*(3), 261–279. DOI: 10.1509/jmkr.45.3.261

Rodríguez-Martínez, G., & Lozano, C. a. A. (2021). Uso de internet y redes sociales en el marco de la contingencia Covid-19 en Colombia. *Texto Livre Linguagem E Tecnologia, 15*, e34828. https://doi.org/DOI: 10.35699/1983-3652.2022.34828

Rogers, E. M. (1995). *Diffusion of Innovations* (4th ed.). The Free Press.

Rogers, E. M. (2003). *Diffusion of innovations* (5th ed.). Free Press.

Romero, D. F. (2021). Digital footprints of #MeToo. www.academia.edu. https://www.academia.edu/114707913/Digital_footprints_of_MeToo

Rosenthal, S. (2022). Information sources, perceived personal experience, and climate change beliefs. *Journal of Environmental Psychology, 81*, 101796. DOI: 10.1016/J.JENVP.2022.101796

Royal Museums Greenwich. (2022). *What can the James Webb Space Telescope do?* https://www.rmg.co.uk/stories/topics/james-webb-space-telescope-vs-hubble-space-telescope

Saba, Z., & Roy, S. (2024). Technological Impact on Tertiary Level Students' Writing Skills in the Post-COVID Era. *In International Journal of Research and Scientific Innovation (IJRSI)*. ISSN No. 2321-2705 | DOI: |Volume XI Issue II February 2024DOI: 10.51244/IJRSI

SABC. (2020). *Social television survey report*. South African Broadcasting Corporation. https://www.sabc.co.za/social-television-survey-report

Sabelhaus, P. A., & Decker, J. E. (2004). An overview of the James Webb Space Telescope (JWST) project. *Optical, Infrared, and Millimeter Space Telescopes, 5487*(12), 550–563. DOI: 10.1117/12.549895

Salaudeen, K. A., & Lawal, H. A. (2019). *Social Media Addiction and Formal Writing Skills Among Mass Communication Students*. Southwestern University.

Salaverría, R. & Cardoso, G. (2023). Future of disinformation studies: emerging research fields. *Profesional De La información Information Professional*, 32(5), 1-7. https://doi.org/DOI: 10.3145/epi.2023.sep.25

Sapar, N., & Hussin, A. R. C. (2024). Role of attitude, norm and behaviour control among young voters in social media toward political engagement. In *Lecture notes on data engineering and communications technologies* (pp. 150–159). https://doi.org/DOI: 10.1007/978-3-031-59711-4_14

Searle, J. R. (1969). *Speech Acts*. Cambridge University Press., https://scholar.google.com/scholar?q=Searle+J.R.+1969:+Speech+Acts&hl=en&as_sdt=0&as_vis=1&oi=scholart

Serem, D. K., Boit, J. M., & Wanyama, M. N. (2013). *Understanding Research; A Simplified Form*. Utafiti Foundation.

Severin, W. J., & Tankard, J. W. (2001). *Communication theories: Origins, methods and uses in the mass media* (5th ed.). Addison Wesley Longman.

Sheth, S. (2019). Super PACs, personal data, and campaign finance loopholes. *Virginia Law Review, 105*(3), 655–702.

Shields, D. C. (2008). *Symbolic Convergence Theory*. Wiley Online Library., DOI: 10.1002/9781405186407.wbiecs125

Short, J., Williams, E., & Christie, B. (1976). *The social psychology of telecommunications*. John Wiley & Sons.

Shukla, S., Singh, P. P., & Garima. (2020). #MeToo movement: Influence of social media engagement on intention to control sexual harassment against women. *Journal of Content, Community & Communication, 12*(6)

Siapera, K. (2012). *The political economy of search engines: An analysis of Google*. Palgrave Macmillan.

SIMElab Africa. (2019). *Social Media Consumption in Kenya*: Practices and Trends. Kenya; USIUAfrica. https://www.usiu.ac.ke/assets/file/SIMElab_Social_Media_Consumption_in_Kenya_report.pdf

Simin, M. T., & Janković, D. (2014). Applicability of Diffusion of Innovation Theory in Organic Agriculture. In *Economics of Agriculture* 2/2014 UDC: 631.147:001.895

Sundar, S. S. (2008). The MAIN model: A heuristic approach to understanding technology effects on credibility. In Metzger, M. J., & Flanagin, A. J. (Eds.), *Digital media, youth and credibility* (pp. 73–100). The MIT Press.

Tajfel, H., & Turner, J. C. (1979). An integrative theory of intergroup conflict. In Austin, W. G., & Worchel, S. (Eds.), *The social psychology of intergroup relations* (pp. 33–47). Brooks/Cole.

Tamilmani, K., Rana, N. P., & Dwivedi, Y. K. (2021). Consumer acceptance and use of information technology: A meta-analytic evaluation of UTAUT2. *Information Systems Frontiers, 23*(4), 987–1005. DOI: 10.1007/s10796-020-10007-6

Tandoc, E. C.Jr. (2021). Fake news. In Tumber, H., & Waisbord, S. (Eds.), *The Routledge companion to media disinformation and populism* (pp. 110–117). Routledge.

Tandoc, E. C.Jr, Lim, D., & Ling, R. (2019). Diffusion of disinformation: How social media users respond to fake news and why. *Journalism, 21*(3), 381–398. DOI: 10.1177/1464884919868325

Tandoc, E. C.Jr, Lim, Z. W., & Ling, R. (2018). Defining "Fake News": A typology of scholarly definitions. *Digital Journalism (Abingdon, England), 6*(2), 137–153. DOI: 10.1080/21670811.2017.1360143

Theocharis, Y., Barberá, P., Fazekas, Z., Popa, S. A., & Parnet, O. (2016). A bad workman blames his tweets: The consequences of citizens' uncivil Twitter use when interacting with party candidates. *Journal of Communication, 66*(6), 1007–1031. DOI: 10.1111/jcom.12259

ThisDayLive. (2021, January 31). *Nigeria's TV industry: A potpourri of opportunities.* https://www.thisdaylive.com/index.php/2021/01/31/nigerias-tv-industry-a-potpourri-of-opportunities/

Thomas, C. L., Ingels, D. J., Kazmi, M. A., Ohu, E. A., Belle, C., & Spitzmueller, C. (2022). Adolescents' problematic internet use in secondary school students in Lagos, Nigeria: A preliminary examination of individual and family-based predictors and consequences. *Computers in Human Behavior, 132*, 107247. DOI: 10.1016/j.chb.2022.107247

TikTok. (2020, January 10). *Brand collaborations with TikTok content creators drive big results*. TikTok for Business. https://www.statista.com/topics/3147/gaming-video-content-market/

Tirado-Morueta, R., Mendoza-Zambrano, D., Marín-Gutiérrez, I., & Mendoza-Zambrano, M. (2017). The relativity of sociodemographic determinism on the digital divide in high school students in Ecuador. *International Journal of Communication, 11*, 1528–1551.

Trindade, L. V. P. (2023). Disability is no laughing matter. A critical discourse analysis of disability mockery on social media in Brazil. *Canadian Journal of Latin American and Caribbean Studies / Revue Canadienne Des Études Latino-américaines Et Caraïbes, 49*(1), 22–36. https://doi.org/ DOI: 10.1080/08263663.2023.2271306

Troisi, O., Fenza, G., Grimaldi, M., & Loia, F. (2022). Covid-19 sentiments in smart cities: The role of technology anxiety before and during the pandemic. *Computers in Human Behavior, 126*, 106986. DOI: 10.1016/J.CHB.2021.106986 PMID: 34511715

Tsoy, D., Tirasawasdichai, T., & Kurpayanidi, K. I. (2021). Role of Social Media in Shaping Public Risk Perception during COVID-19 Pandemic: A Theoretical Review. *International Journal of Management Science and Business Administration*, *7*(2), 35–41. DOI: 10.18775/IJMSBA.1849-5664-5419.2014.72.1005

Tuters, M. (2020). Fake news and the Dutch YouTube political debate space. In Rogers, R., & Niederer, S. (Eds.), *The Politics of Social Media Manipulation* (pp. 218–238). Amsterdam University Press.

Twitch. (2020, January 31). *Co-streaming on Twitch.* Twitch. https://help.twitch.tv/s/article/twitch-content-sharing-guidelines?language=en_US

Twitter. (2020, October 22). *How Twitter is preparing for the 2020 US election.* Twitter Blog. https://www.frontiersin.org/articles/10.3389/fsoc.2019.00079/full

Valerie, M., & Campbell, V. M. (2020). Digital Stories as Data. In Servaes, J. (Ed.), *Handbook of Communication for Development and Social Change*. Springer.

Valkenburg, P. M., & Peter, J. (2013). The differential susceptibility to media effects model. *Journal of Communication*, *63*(2), 221–243.

Van den Broeck, A., Ferris, D. L., Chang, C.-H., & Rosen, C. C. (2016). A review of self-determination theory's basic psychological needs at work. *Journal of Management*, *42*(5), 1195–1229. DOI: 10.1177/0149206316632058

van der Linden, S., Panagopoulos, C., & Roozenbeek, J. (2020). You are fake news: Political bias in perceptions of fake news. *Media Culture & Society*, *42*(3), 460–470. DOI: 10.1177/0163443720906992

van Doorn, J., Lemon, K. N., Mittal, V., Nass, S., Pick, D., Pirner, P., & Verhoef, P. C. (2010). Customer engagement behaviour: Theoretical foundations and research directions. *Journal of Service Research*, *13*(3), 253–266. DOI: 10.1177/1094670510375599

Van Gils, F., Müller, W., & Prüfer, J. (2024). Microtargeting, voters' unawareness, and democracy. *Journal of Law Economics and Organization*. Advance online publication. DOI: 10.1093/jleo/ewae002

Venkatesh, V., Morris, M. G., Davis, G. B., & Davis, F. D. (2003). User acceptance of information technology: Toward a unified view. *Management Information Systems Quarterly*, *27*(3), 425–478. DOI: 10.2307/30036540

Venkatesh, V., Thong, J. Y., & Xu, X. (2012). Consumer acceptance and use of information technology: Extending the unified theory of acceptance and use of technology. *Management Information Systems Quarterly*, *36*(1), 157–178. DOI: 10.2307/41410412

Wagner, M. C., & Boczkowski, P. J. (2019). The reception of fake news: The interpretations and practices that shape the consumption of perceived misinformation. *Digital Journalism (Abingdon, England)*, *7*(7), 870–885. DOI: 10.1080/21670811.2019.1653208

Wahutu, J. S. (2019). Fake News and Journalistic "Rules of the Game". *African Journalism Studies*, *40*(4), 13–26. DOI: 10.1080/23743670.2019.1628794

Walther, J. B. (1992). Interpersonal effects in computer-mediated interaction: A relational perspective. *Communication Research*, *19*(1), 52–90.

Wamalwa, I. W., Mburu, B. K., & Mang'uriu, D. G. (2016). Agro Climate and Weather Information Dissemination and Its Influence on Adoption of Climate Smart Practices among Small Scale Farmers of Kisii County, Kenya. In *Journal of Biology, Agriculture and Healthcare*. ISSN 2224-3208 (Paper) ISSN 2225-093X (Online), Vol.6, No.10, 2016

Warner, J. (2016). Adolescents' Dialogic Composing With Mobile Phones. *Journal of Literacy Research*, *48*(2), 164–191. DOI: 10.1177/1086296X16660655

Wicker, A. W., & Doyle, J. (2012). Leveraging multiple mechanisms for information propagation. In *Lecture notes in computer science* (pp. 1–2). https://doi.org/DOI: 10.1007/978-3-642-27216-5_1

Wil, C. S. C., Yunus, M. M., & Suliman, A. (2019). The Use of Social Media to Assist Writing Skills among Secondary Pupils. *International Journal of Academic Research in Progressive Education and Development, 8*(3), 224–236.

Willis, S., & Tranter, B. (2006). Beyond the 'digital divide.'. *Journal of Sociology (Melbourne, Vic.), 42*(1), 43–59. DOI: 10.1177/1440783306061352

Wimmer, D., & Dominick, R. (2011). *Mass media research: An introduction*. California: Belmont.

Winter, S., Neubaum, G., Eimler, S. C., Gordon, V., Theil, J., Herrmann, J., Meinert, J., & Krämer, N. C. (2020). How to cope with challenging tasks: The role of live chats and comments in social TV. *Computers in Human Behavior, 102*, 146–156. DOI: 10.1016/j.chb.2019.08.023

Wood, A. K., & Ravel, A. M. (2018). Fool me once: Regulating "fake news" and other online advertising. *Southern California Law Review, 91*(6), 1223–1278.

Wright, G. S., Reike, G., Barella, P., Boeker, T., Colina, L., van Dishoeck, E., Driggers, P., Goodson, G., Greene, T., Heske, A., Henning, T., Lagage, P.-O., Meixner, M., Norgaard-Nielsen, H., Olofsson, G., Ray, T., Ressler, M., Thatcher, J., Waelkens, C., & Zehnder, A. (2008). Design and development of MIRI, the mid-IR instrument for JWST. *Space Telescopes and Instrumentation 2008. Optical, Infrared, and Millimeter, 7010*, 269–278. DOI: 10.1117/12.790101

YouTube. (2020, October 27). *How YouTube supports elections*. YouTube Official Blog. https://link.springer.com/article/10.1007/s42087-020-00156-0

Zillich, A. F. (2020). Socially shared television viewing: Preconditions, processes and effects of co-viewing and social TV. In Katzenbach, C., & Schrape, J. T. M. (Eds.), *Transforming communications: Studies in cross-media research* (pp. 117–136). Palgrave Macmillan.

About the Contributors

Alfred Akwala graduated with a bachelors degree in Linguistics and Literature in 2000. In 2008 graduated with a masters degree in Linguistics and in 2015 graduated with a PhD in Communication studies Moi University, Department of Communication. Currently Lecturer Technical University of Kenya-Nairobi(Kenya).

***

Adebowale Adetayo is an academic staff of Babcock University. His research interest is Library Science, Social media, Knowledge Management, and Business Information Management. He has published many articles in reputable journals and currently working on projects relating to pandemics, vaccines and virtual learning. He is a graduate of Babcock University.

Romica Bhat is a seasoned academician with an impressive track record of 13 years in academia and an additional 2 years in the industry. Presently, she holds the esteemed position of Associate Professor and Head of the Department of Amity School of Communication, Kolkata. Her areas of expertise encompass Broadcast Journalism and Digital Marketing, fields in which she has demonstrated her prowess through more than 10 published research papers in esteemed national and international journals. Dr. Bhat's academic prowess extends beyond research, as she has had the privilege of presenting her scholarly works at renowned institutions such as IIT Roorkee and Bhartiya Vidyapeeth. She boasts comprehensive proficiency in the key software employed in the Mass Communication industry and possesses a wealth of practical experience, having successfully managed over 500 radio and television projects. Dr. Bhat's impact extends beyond the classroom, as she serves as a valued member of the Board of Studies in prestigious private institutions and holds experience in dealing with the accreditation systems of NAAC and NIRF. Her acumen in organizing academic events is evident through her successful handling of 3 Media conclaves and 2 National Seminars at the university level. Dr. Bhat's dedication to bridging the gap between academia and industry is showcased through her regular conduction of industry-academia sessions within the university. Additionally, she has been entrusted with overseeing departmental admissions at various universities for a decade, a testament to her administrative expertise. In summary, Dr. Romica Bhat's illustrious career in academia, research contributions, practical experiences, and organizational skills position her as a remarkable and accomplished professional in the realm of Mass Communication.

Sayantan Ghosh teaches Journalism at Xavier's College, (autonomous) Kolkata. He is a political columnist with several news organisations, including The Quint and News18.com. He also writes a fortnightly column named Here, There, Everywhere at the Free Press Journal newspaper. He has been a journalist for over seven years and is also a former policy research fellow at the Delhi Assembly Research Centre. He is currently pursuing his Ph.D. in journalism and mass communication.

Ruben Arnoldo Gonzalez is a journalist and scholar. He holds a PhD in Communication Studies from the University of Leeds (UK). He has been reporter and editor of the newspaper La Voz de Michoacán (Morelia, Mexico). He is currently a full-time professor and researcher at the Institute of Government Sciences and Strategic Development (Benemérita Universidad Autónoma de Puebla, Mexico), where he chaired the Centre for Political Communication Studies (2019-2023). He is a member of the National Researchers System (Level 1) of the National Council for Science, Technology, and Humanities (Mexico). He was appointed as Mexico Chair 2022 at Temple University (Philadelphia, USA), where he was a visiting scholar at the Klein College of Media and Communication.

His research interests are journalism studies (mainly professionalism and anti-press violence), political communication, and media systems in emergent democracies.

Adamkolo Mohammed Ibrahim is Lecturer, Department of Mass Communication, University of Maiduguri, Nigeria. He holds a PhD (Bayero University, 2025), an MSc (Universiti Putra Malaysia, 2017) and a BA (Hons) in Mass Communication. His research addresses social television, OTT/streaming media, ICT adoption, fake news, media literacy and development communication, applying UTAUT2 and culturally grounded frameworks. He has authored 100+ peer-reviewed articles, chapters and monographs and led TETFund fieldwork (300 survey respondents; six FGDs) on misinformation and national integration. He designed curricula, provides copy-editing and communications consultancy, and has been recognised with a Publons Global Peer Review Award (Top 1%) and AD Scientific Index citation rankings.

Grace Mwanja is a Communications Professional with 15 years of experience..Keen on improving public communications to shape perceptions for peaceful coexistence in the society.

Sreethi Rebeka R is a PhD Research Scholar at Dept. of Commerce, Christ University, Bengaluru.

Ranjit Singha is a Doctorate Research Fellow at Christ (Deemed to be University) and a distinguished American Psychological Association (APA) member. His expertise lies in research and development across various domains, including Mindfulness, Addiction Psychology, Women Empowerment, UN Sustainable Development Goals, and Data Science. He has earned certifications from renowned institutions, including IBM and The University of Oxford Mindfulness Centre, UK, in Mindfulness. Additionally, he holds certifications as a Microsoft Innovative Educator, Licensed Yoga Professional, Certified Mindfulness Teacher, and CBCT Teachers Training from Emory University, USA. Mr Ranjit's educational qualifications include PGDBA (GM), MBA (IB), MSc in Counseling Psychology, and completion of a Senior Diploma in Tabla (Musical Instrumentation). His dedication to continuous learning is evident through his involvement in the SEE Learning® (Social, Emotional, and Ethical) Learning program. As a committed researcher and educator, Mr Ranjit focuses on mindfulness and compassion-based interventions. He has an impressive publication record, having authored twenty-three research papers, ten chapters, four books, and five edited books. His research interests encompass various aspects of mindfulness, such as assessment, benefits of mindfulness-based programs, change mechanisms, professional training, mindfulness ethics, cognitive and neuropsychology, and studies related to high-risk behaviours. Apart from his research endeavours, Mr Ranjit has extensive teaching experience, instructing courses in diverse subjects like Forensic Psychology, Positive Psychology, Organizational Planning, Strategic Management, Psycho Metric Tests, Counseling Skills, Disaster Management, Basic Computer Science, Business Planning, Business Law, and Auditing. He has mentored numerous Postgraduate and undergraduate research projects, demonstrating his commitment to nurturing young minds in psychology. Furthermore, Mr Ranjit actively provides personal counselling services, showcasing his genuine concern for his students' well-being and academic success. His unwavering dedication to research and education has solidified his position as a valuable contributor to psychology.

Surjit Singha is a passionate academician and researcher with over 16 years of professional experience spanning administration, teaching, and research. Currently serving as a faculty member at Kristu Jayanti (Deemed to be University), Dr. Singha is deeply committed to fostering academic growth and guiding students in their research pursuits. His multidisciplinary interests include the UN Sustainable Development Goals, Organizational Climate, Workforce Diversity, Organizational Culture, Human Resource Management (HRM), Marketing, Finance, International Business, Global Business, Artificial Intelligence, K12 & Higher Education, Gender, and Cultural Studies. An accomplished author and editor, Dr. Singha contributes to renowned global publications and journals and holds prominent editorial roles, including being on the Editorial Review Board of the Information Resources Management Journal (SCOPUS & WoS). With an impressive academic portfolio, Dr. Singha has published over 160+ articles, book chapters, and textbooks, including 100+ SCOPUS-indexed publications. He also holds 2+ US Copyrights and 1+ Patent, demonstrating his dedication to innovation and originality. Additionally, he has successfully completed two fully funded minor research projects from Kristu Jayanti (Deemed to be University). As a research mentor, Dr. Singha has guided numerous students and young scholars toward academic excellence. His academic qualifications include a PhD, M.Phil, and MBA (Distinction) from Christ University, M.Com, BBM from Kristu Jayanti College (Bangalore University), and PGDHE (Higher Education) from IGNOU (Central University). He is also a UGC-NET (Commerce) qualifier and serves as an Examiner for International Baccalaureate (IB).

Index

www.ingramcontent.com/pod-product-compliance
Ingram Content Group UK Ltd.
Pitfield, Milton Keynes, MK11 3LW, UK
UKHW050946290926
15470UKWH00005B/1666